T0205375

Lecture Notes in Networks and Systems 942

The series "Lecture Notes in Networks and Systems" publishes the latest developments in Networks and Systems—quickly, informally and with high quality. Original research reported in proceedings and post-proceedings represents the core of LNNS.

Volumes published in LNNS embrace all aspects and subfields of, as well as new challenges in, Networks and Systems.

The series contains proceedings and edited volumes in systems and networks, spanning the areas of Cyber-Physical Systems, Autonomous Systems, Sensor Networks, Control Systems, Energy Systems, Automotive Systems, Biological Systems, Vehicular Networking and Connected Vehicles, Aerospace Systems, Automation, Manufacturing, Smart Grids, Nonlinear Systems, Power Systems, Robotics, Social Systems, Economic Systems and other. Of particular value to both the contributors and the readership are the short publication timeframe and the world-wide distribution and exposure which enable both a wide and rapid dissemination of research output.

The series covers the theory, applications, and perspectives on the state of the art and future developments relevant to systems and networks, decision making, control, complex processes and related areas, as embedded in the fields of interdisciplinary and applied sciences, engineering, computer science, physics, economics, social, and life sciences, as well as the paradigms and methodologies behind them.

Indexed by SCOPUS, INSPEC, WTI Frankfurt eG, zbMATH, SCImago.

All books published in the series are submitted for consideration in Web of Science.

For proposals from Asia please contact Aninda Bose (aninda.bose@springer.com).

Arthur Gibadullin
Editor

Digital and Information Technologies in Economics and Management

Proceedings of the International Scientific and Practical Conference “Digital and Information Technologies in Economics and Management” (DITEM2023)

Editor
Arthur Gibadullin
National Research University "Moscow Power Engineering Institute"
Moscow, Russia

ISSN 2367-3370 ISSN 2367-3389 (electronic)
Lecture Notes in Networks and Systems
ISBN 978-3-031-55348-6 ISBN 978-3-031-55349-3 (eBook)
https://doi.org/10.1007/978-3-031-55349-3

This Springer imprint is published by the registered company Springer Nature Switzerland AG
The registered company address is: Gewerbestrasse 11, 6330 Cham, Switzerland

Paper in this product is recyclable.

Preface

The conference was held with the aim of summarizing international experience in the field of digital development of the economy and management, the introduction of information technologies and systems in the organizational processes of managing corporations and individual industries.

The III International Scientific and Practical Conference "Digital and Information Technologies in Economics and Management" (DITEM2023) was held on November 21–23, 2023.

The conference addressed issues of networks and systems related to the use of information technologies in economics, management and management of various sectors. A distinctive feature of the conference is that it featured presentations by authors from China, Bulgaria, Uzbekistan, Oman, Kazakhstan and Russia. Researchers from different countries presented the process of transition to new information technologies of various network and system structures and sectors.

The conference sessions were moderated by Gibadullin Arthur of the National Research University "Moscow Power Engineering Institute", Moscow, Russia and Manuchehr Sadriddinov of the International University of Tourism and Entrepreneurship of Tajikistan, Dushanbe, Tajikistan.

Thus, the conference made it possible to develop new scientific recommendations on the use of information, computer, digital and intellectual technologies and networks in industry and fields of activity that can be useful to state and regional authorities, international and supranational organizations, the scientific and professional community.

Each presented paper was reviewed by at least three members of program committee or an independent reviewer. As a result of the work of all reviewers, 21 papers were accepted for publication out of the 53 received submissions. The reviews were based on the assessment of the topic of the submitted materials, the relevance of the study, the scientific significance and novelty, the quality of the materials and the originality of the work. Reviewers, program committee members, and organizing committee members did not enter into discussions with the authors of the articles.

The organizing committee of the conference expresses its gratitude to the staff at Springer who supported the publication of this proceeding. In addition, the organizing committee would like to thank the conference participants, reviewers, and everyone who helped organize this conference and shape the present volume for publication in the Springer LNNS series.

Arthur Gibadullin
Manuchehr Sadriddinov

Organization

Program Committee Chairs

Asrorzoda Ubaydullo	International University of Tourism and Entrepreneurship of Tajikistan, Tajikistan
Sadriddinov Manuchehr	International University of Tourism and Entrepreneurship of Tajikistan, Tajikistan
Gibadullin Arthur	National Research University "Moscow Power Engineering Institute," Russia

Program Committee

Firsov Yury	Prague Institute for Advanced Studies, Czech Republic
Łakomiak Aleksandra	Wroclaw University of Economics and Business, Poland
Davlatov Davlatmakhmad	Mining and Metallurgical Institute of Tajikistan, Tajikistan
Sadullozoda Shahriyor	Khujand Polytechnic Institute of Tajik Technical University named after academician M.S.Osimi, Tajikistan
Tao Itao	Shenzhen University, China
Aikenova Ryskeldy	University "TURAN-ASTANA", Kazakhstan
Wang Yanming	Shenzhen University, China
Dimitrov Lubomir	Technical University of Sofia, Bulgaria
Koryachko Marina	Moscow Polytechnic University, Russia
Karapetkov Stanimir	Technical University of Sofia, Bulgaria
Britvina Valentina	Moscow Polytechnic University, Russia
Geetha Devi	National University of Science & Technology, Oman
Khalmatjanova Gulchekhra	Fergana State University, Fergana, Uzbekistan
Raviprakash Dani	Texas Tech University, USA
Sherovna Guzal	Fergana State University, Fergana, Uzbekistan

Organizing Committee

Gibadullin Artur	National Research University "Moscow Power Engineering Institute," Russia
Voinash Sergey	Kazan Federal University, Russia
Sadriddinov Manuchehr	International University of Tourism and Entrepreneurship of Tajikistan, Tajikistan
Kamenova Mazken	University "TURAN-ASTANA," Kazakhstan

Organizer

International University of Tourism and Entrepreneurship of Tajikistan, Tajikistan

Contents

Development and Application of Information Systems for Planning Production Activities of an Industrial Company

M. Yu. Ivanov(✉) and V. V. Lobova

Bratsk State University, 40 Makarenko, Bratsk 665709, Russia
libmann.52@gmail.com

Abstract. The growth of production efficiency is one of the key priorities of economic development. Improving product quality, reducing production costs, expanding the range of manufactured products are the main strategic goal of any enterprise. Business entities cannot operate and develop without effective management systems that utilize the latest information technologies. Constantly changing market requirements and huge data flows require managers to make quick and efficient decisions aimed at maximizing profits at minimal costs. The article presents the results of research of theoretical prerequisites of digitalization of resource planning processes of industrial organizations, contributing to the reduction of the real cost of production and increasing the productivity of the metallurgical plant. The operational activity of the enterprise is formalized in the form of a certain enlarged business process, within which groups of process indicators corresponding to those or other constituent elements of the process are allocated. A computer program has been developed to optimize and improve the quality of the process of calculating the main technical and economic parameters of the foundry's operation, as well as to obtain statistical reports on the performed operations. Unlike typical foreign and domestic solutions, the application takes into account the specifics of metallurgical enterprises, does not contain redundant functionality, does not require computer resources, can be used by employees of planning and finance and planning and technical departments, as well as in the training of specialists.

Keywords: Enterprise Management · Resource Planning · Accounting · Manufacturing · Non-Ferrous Metallurgy · Foundry · Automation

1 Introduction

Nowadays, cost optimization, improvement of production efficiency in accordance with the ever-increasing requirements of consumers in a highly competitive environment cannot be based only on the intuition of even the most experienced managers.

Operational control over all costs of the organization, mathematical methods of analysis, forecasting and planning based on the consideration of many parameters and criteria are necessary [1].

A. Gibadullin (Ed.): DITEM 2023, LNNS 942, pp. 1–13, 2024.
https://doi.org/10.1007/978-3-031-55349-3_1

Modern materials planning is a tightly integrated system that spans the entire plant and tracks all operations, constantly making changes to production schedules to ensure the company is running smoothly to meet customer demand and expectations.

Improving the quality of management decisions and competitiveness of the enterprise in dynamic markets, in turn, is impossible without the use of modern information technologies [2].

2 Scientific Novelty

The scientific novelty of the research lies in the development of information systems and technologies of enterprise management, providing a guarantee of availability of the necessary quantity of required materials and components along with a possible reduction of stocks and, consequently, a reduction in storage costs, minimization of equipment downtime.

3 Materials and Research Methods

System analysis to find out the causes of existing difficulties, setting goals, developing options for solving problems of planning the activities of the enterprise; economic analysis to forecast the results of production; high-level programming methods to develop an application in the programming language «Borland Delphi».

4 Theoretical Prerequisites for Digitalization of Resource Planning Processes

It is known that the main goal of industrial organizations is to effectively manage the inventory of raw materials, supplies and equipment [3].

Traditional planning calculates material requirements (production and purchase orders). This uses an infinite or unlimited capacity model, i.e. only materials are considered and ignores capacity problems or constraints.

However, there are companies with limited production capacity and planning based on the unlimited capacity model is not suitable for them. Limitations may be, for example, production resources such as smelting furnaces or drying rooms, the amount of equipment, specialized specialists, and so on.

During operation, raw material volumes need to be checked against available equipment using a separate capacity planning tool. This is a step-by-step two-step process that can be time-consuming. Yes, it is a workable solution and a huge step in production planning, but automated advanced planning systems take into account both material and capacity data, allowing you to create production plans tailored to your constraints.

The automation of these processes allows optimal regulation of deliveries, controlling stock levels and the production technology itself.

In view of the above, digital solutions for managing a manufacturing enterprise realize the following functions:

- description of raw materials, materials and composition of finished products;

- inventory management;
- compilation of a materials inventory;
- calculating the need for materials and components;
- production equipment accounting;
- accounting of production operations;
- calculating the requirements for necessary equipment and personnel;
- formation of the production program;
- determination of the cost of manufactured products and the possibility of reducing production costs [4, 5].

Manufacturing resource planning systems are based on the management of multiple processes, each of which is interconnected: strategic business planning, production planning, sales and operations planning, master production scheduling (see Table 1), materials planning (see Table 2), capacity planning, and capacity and materials execution support systems. The results of the interaction of these processes are integrated with various financial documents: business plan, purchasing report, shipment plan and inventory forecast in value terms.

Table 1. Calendar planning.

Planning level	Object	Planning horizon	Planning interval	Evaluation of implementation
Sales and operations plan	Commodity group	1–2 years	Quarter or month	Quarterly
Master production schedule	Products of independent demand and schedule final assemblies	Quarter-year	A month or a week	Monthly
Material requirement plan	Dependent demand products	1–6 months	A week or a day	Weekly
Operational management of production	Technological Operations	1–4 weeks	Day or hour	Daily

Operational production management is intended to form schedules for the fulfillment of production orders in terms of technological operations and represents the most detailed plan. It makes sense to develop it within the planning horizon, within which the plan has already been confirmed for execution and its changes are unlikely.

Modern conditions of functioning of metallurgical enterprises require actual ways of modernization of existing planning and management systems through the development of automated solutions, which will allow to achieve long-term strategic goals.

To date, in the field of metallurgical production, negative trends of falling production and sales remain, which is the cause of inefficient resource allocation, imperfect management systems and irrationality of decisions [6].

Table 2. Resource planning.

Stage	Plan	Object	Input data	Weekend data
Planning for resource requirements	Sales and operations plan	Product group	Production plan, resource profile	Resource requirements
Aggregated capacity requirement planning	Master calendar	Independent demand products	Available resources, master production schedule, resource utilization profile	Aggregated capacity requirements
Capacity requirement planning	Material requirement plan	Components and materials	Material requirement plan	Capacity requirement plan

Modern software can help in solving the arising problems, which will allow the enterprise to increase its competitiveness, reduce the degree of impact of negative external and internal factors on the financial and economic activities, which, thus, will contribute to increasing profits and strengthening financial stability.

In recent years, one of the most popular and widely used tools used in enterprises to realize strategic objectives is the balanced scorecard. It allows not only to identify the key performance indicators of the organization in four main areas: finances, customers, processes, prospects, but also to control the process of achieving the plan by setting target values of these indicators - both generally accepted norms and values determined within a particular business entity [7].

To reduce the negative impact of the above-mentioned factors, it will be possible to build a balanced system based on the application of process indicators rather than on the classical approach of developing performance indicators in four areas.

Let's represent the operational activity of the enterprise in the form of a certain enlarged business process, within which we will allocate groups of process indicators corresponding to those or other constituent elements of the process (see Fig. 1).

Figure one shows the groups of process indicators: the first - indicators characterizing the supply of raw materials; the second - characterizing the production process, the third - characterizing customer satisfaction and the fourth group - indicators characterizing the results of the production process.

In such a division of process indicators into groups, similar to the groups of performance indicators of the classical balanced scorecard, there is a clear relationship and hierarchy. Compliance with the set criteria at the stage of supply of raw materials and supplies will give the corresponding production results and provides coherence, continuity and efficiency of the production process, which positively affects customer satisfaction, the conquest of new markets and, subsequently, favorably affects the financial results of the operating activity of the enterprise [8].

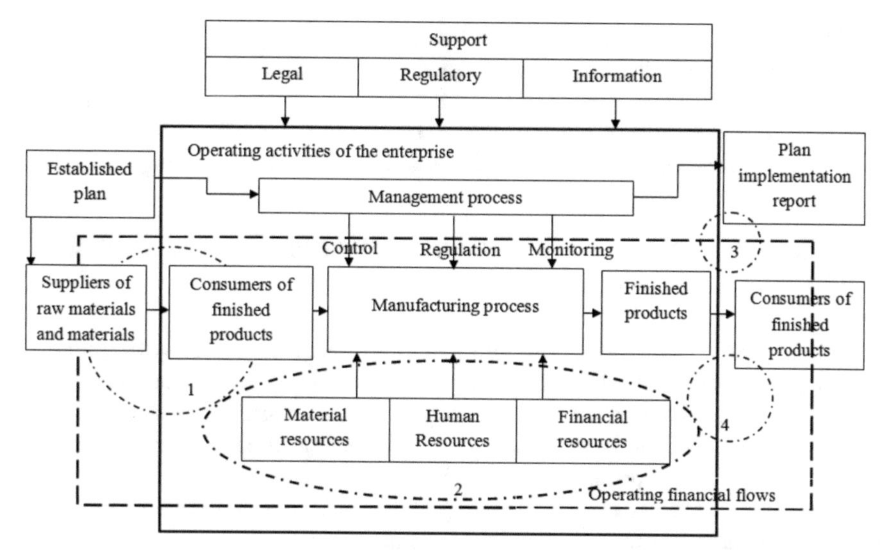

Fig. 1. Business process "Operating activities of the enterprise".

A clear understanding of the processes that take place within the operational activities of the enterprise and ensure continuous production and realization of products, structured stages allow to achieve greater efficiency and effectiveness in the selection of key performance indicators when building a balanced management system. The use of the process approach allows to mutually link the production process, which is a key aspect for industrial enterprises, and the process of developing tools for the realization of strategic goals. It makes the balanced scorecard more flexible, compliant with the requirements and understandable for all stakeholders. This relationship allows avoiding the situation of duplication of indicators or inefficient borrowing of previously developed, but not suitable for this enterprise, management systems [9].

The target values of performance indicators are set by each enterprise individually, as it is impossible to apply a uniform approach in this case. The target values depend on the financial situation, the chosen policy of financing activities, the stage of development of the enterprise and so on.

In view of the above, the development of independent solutions for the digitalization of management and planning activities of an industrial enterprise is economically feasible, and their implementation and operation are cost-effective and profitable. Saving the working time fund of specialists creates prerequisites for reengineering of business processes related to the activities of planning and technical structural units of the organization.

For business entities there is no need to purchase multi-branch sets of software and target configurations with additional payment for information and service maintenance, maintenance and support.

In addition, third-party software products require labor-intensive, time-consuming and not always successful customization and adaptation to the needs of a particular enterprise department and activity.

The social effect of the developments lies in the possibility of automated application of mathematical methods for solving managerial tasks and reducing the influence of the human factor on the correctness of the latter; partial release of workers from routine operations; minimization of the probability of error in the course of data processing; reduction of paperwork, which, in turn, increases the comfort of working conditions.

5 Practical Implementation of an Application for Automated Planning of Foundry Work

Automation systems for planning processes are intended to improve production efficiency. They can be used to solve a large number of production issues that arise.

Modern manufacturing consists not only of machines and equipment, but also of software that helps in purchasing raw materials, invoicing, and forecasting the required amount of materials for production [10, 11].

The planning process is an important part of continuous and efficient production, which is why many businesses pay a lot of attention to it.

Foundry production of aluminum alloys and related products is constantly increasing due to the significant growth in demand for them. Analysis of the foundry industry shows that the scale of aluminum consumption has a direct relationship with the general level of development of the country's economy. Also, the volume of aluminum consumption is closely related to the standard of living of the population.

The main end users of aluminum alloys are machine-building industries producing various vehicles (automobiles, trains and cars, airplanes, ships); construction; production of various types of soft and hard packaging materials, as well as electrical equipment.

However, foundry production of aluminum alloys is associated with increasing costs of fuel and energy, environmental protection and, as a consequence, leads to reduced efficiency and low profitability of metallurgical enterprises. In order to be competitive in the aluminum alloys market and win leading positions, it is necessary to radically improve the efficiency and profitability of foundries, including through the development and application of modern information systems.

The figure shows the modules of the production resource and process planning system being developed (see Fig. 2).

Sales and operations planning is the key link between the strategic planning and business planning process and the detailed planning and execution system of the enterprise plan. Moreover, the sales and operations plan is the basis for all other plans and schedules.

The demand management unit links the functions of demand forecasting, customer order handling, distribution and material movement between production sites.

The master production schedule describes planning based on nomenclature units of independent demand, i.e. what, when and how much to produce. It is developed from the sales and operations plan and is the basis for all other plans.

Material requirement planning is a calculation mechanism that calculates the need for materials (in all nomenclature items) that do not represent products of independent demand. The need for such items is calculated on the basis of demand data for products sold by the company.

The stock operations block keeps the stock data of nomenclature items up to date.

The block of planned receipts for open orders is required to work with production and purchase orders.

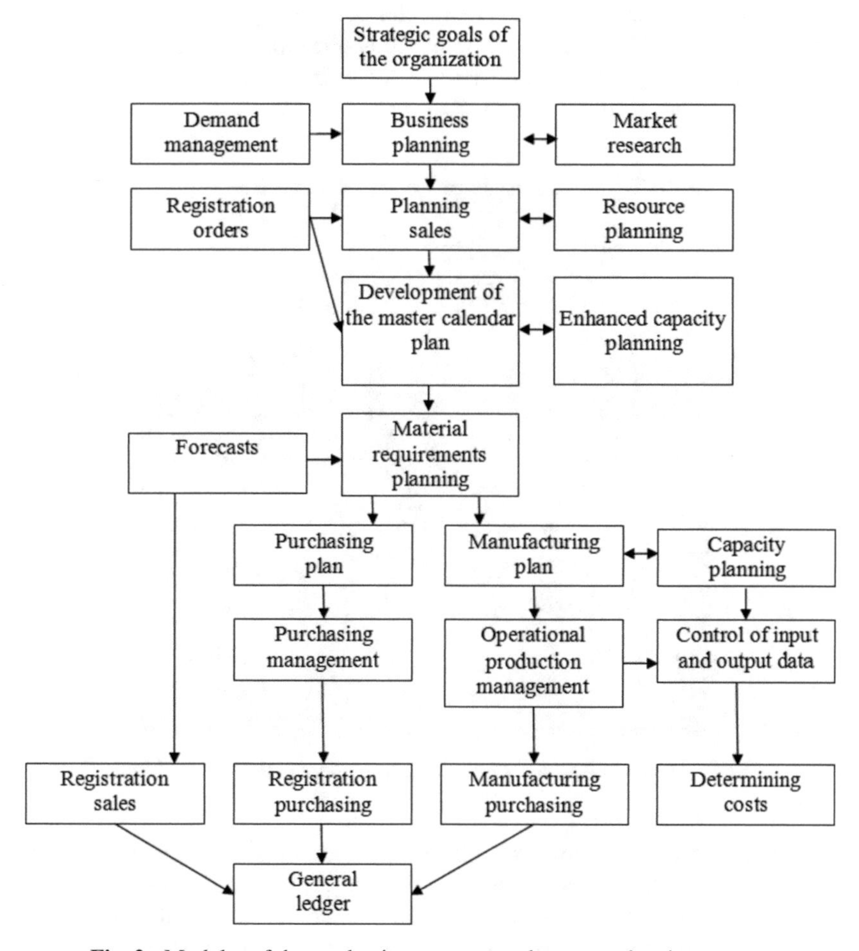

Fig. 2. Modules of the production resource and process planning system.

Capacity demand planning provides information on work center utilization according to the program adopted at the master schedule level.

Management of input or output material flows allows you to control the execution of the capacity utilization plan.

Supply management monitors the implementation of the procurement plan.

Distributed resource planning provides planning when an enterprise has a geographically distributed structure.

The logic of functioning of the projected application is based on the example of functioning of a business entity of the real sector of economy - a foundry.

The next step in the implementation of the application is to create the necessary button forms in the programming language "Borland Delphi". Thus, the main form and the «Tables» form are used to move to other forms (see Fig. 3).

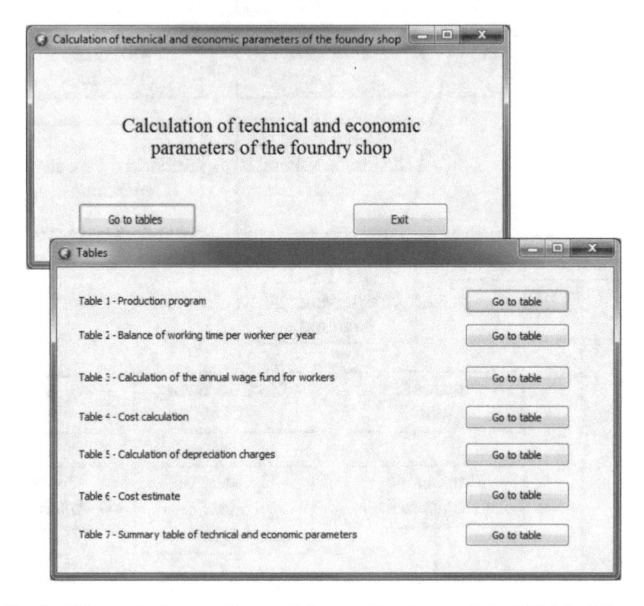

Fig. 3. The main button form of the application and the "Tables" form.

The forms "Production program", "Balance of working time per worker per year", "Calculation of the annual wage fund for workers", "Cost calculation", "Calculation of depreciation charges", "Cost estimate" are intended for input of initial data and calculation of basic technical and economic parameters of foundry shop operation.

For the convenience of user's perception of information, some button forms of the application ("Balance of working time per worker per year", "Calculation of the annual wage fund for workers", "Cost Calculation", "Calculation of depreciation charges") contain formulas explaining the mechanism of calculation of production indicators (see Figs. 4 and 5).

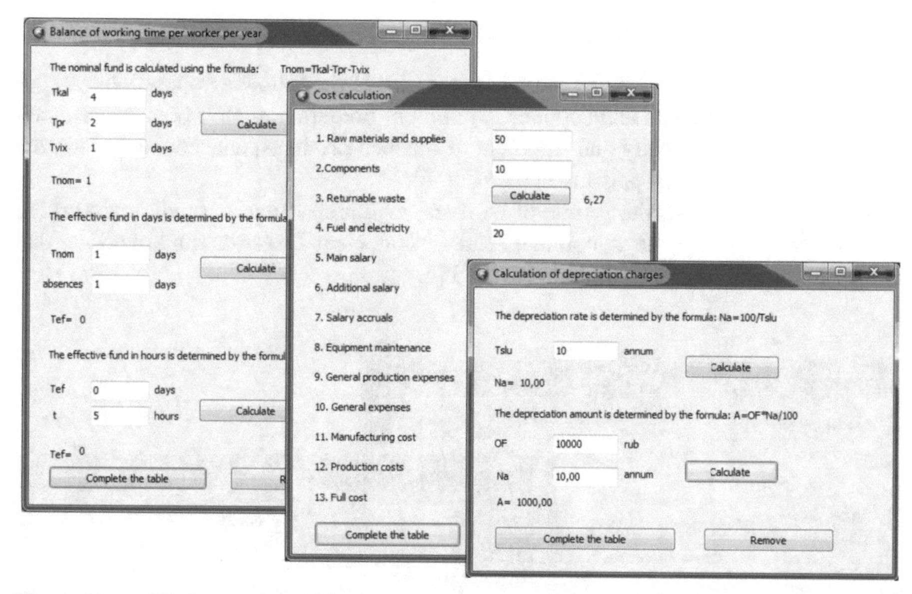

Fig. 4. Forms "Balance of working time per worker per year", "Cost calculation", "Calculation of depreciation charges".

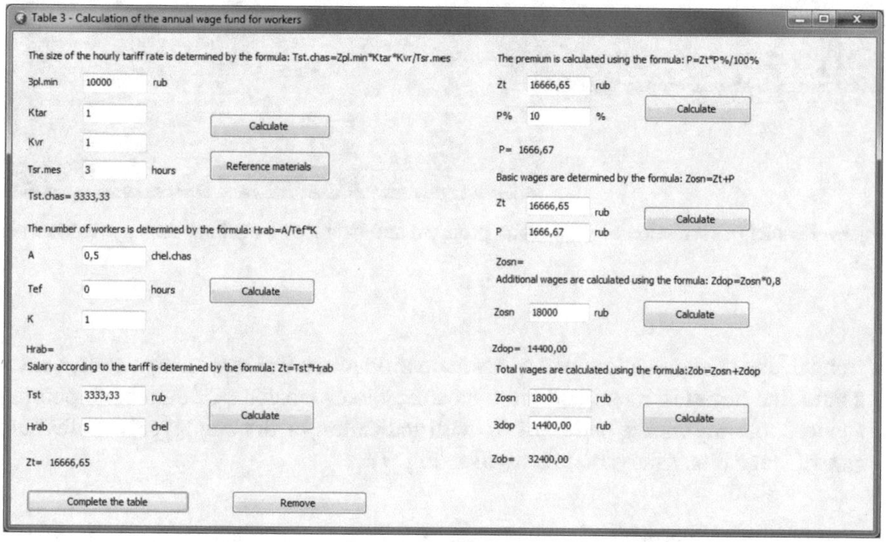

Table 3 - Calculation of the annual wage fund for workers

The size of the hourly tariff rate is determined by the formula: Tst.chas=Zpl.min*Ktar*Kvr/Tsr.mes
3pl.min 10000 rub
Ktar 1
Kvr 1
Tsr.mes 3 hours
Calculate
Reference materials
Tst.chas= 3333,33
The number of workers is determined by the formula: Hrab=A/Tef*K
A 0,5 chel.chas
Tef 0 hours
K 1
Hrab=
Salary according to the tariff is determined by the formula: Zt=Tst*Hrab
Tst 3333,33 rub
Hrab 5 chel
Zt= 16666,65
Complete the table
Remove

The premium is calculated using the formula: P=Zt*P%/100%
Zt 16666,65 rub
P% 10 %
P= 1666,67
Basic wages are determined by the formula: Zosn=Zt+P
Zt 16666,65 rub
P 1666,67 rub
Zosn=
Additional wages are calculated using the formula: Zdop=Zosn*0,8
Zosn 18000 rub
Zdop= 14400,00
Total wages are calculated using the formula:Zob=Zosn+Zdop
Zosn 18000 rub
3dop 14400,00 rub
Zob= 32400,00

Fig. 5. Form "Annual wage fund for workers".

Estimated technical and economic parameters of foundry operation are:

- production program, showing the labor intensity of the following operations: acceptance of raw materials and supplies, equipment preparation, melt preparation, melt fluxing, casting, delivery and shipment of finished products, unaccounted work, the entire volume of work performed (see Fig. 6);
- annual balance of working time of workers, including calendar fund, non-working (holiday, weekends) days, nominal fund, absenteeism, effective fund, working day length, effective working time fund (see Fig. 6);

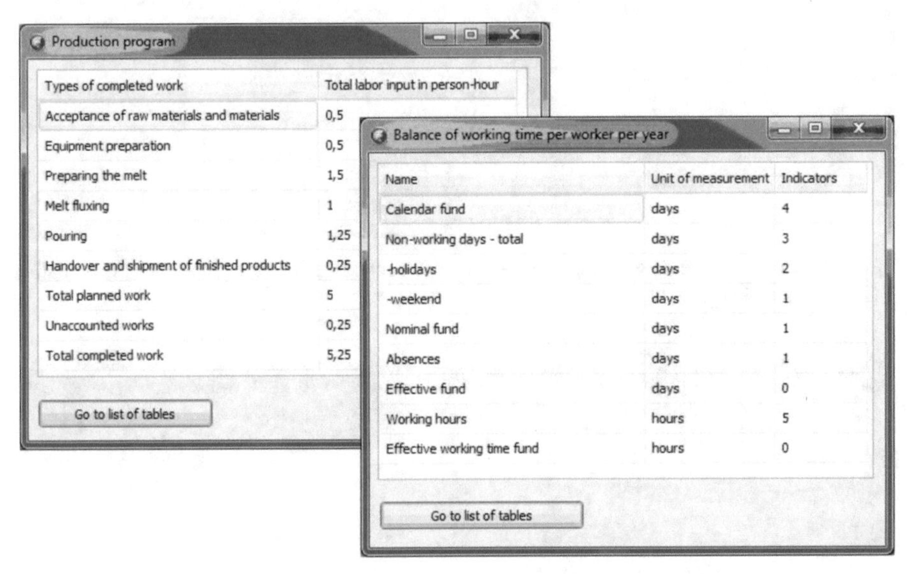

Fig. 6. Results of formation of production program and balance of working time per worker per year.

- annual labor remuneration fund of the main production workers (charge maker, automatic line adjuster, metal and alloy smelter, fluxer, operator of automatic pouring devices, casting picker and marker) with indication of the employee's profession, grade, wage rate, salary, bonus, etc. (see Fig. 7);
- cost of production, which includes expenses for raw materials, components, waste, fuel and electricity, wages of the main production workers, payroll accruals, maintenance of equipment, general production expenses, general business expenses and other production costs (see Fig. 8);
- depreciation charges taking into account the type and quantity of fixed production assets (buildings and structures, mixer, conveyor, casting machine, metal cutting line, rolling mill), service life and cost of equipment, depreciation rate (see Fig. 9).

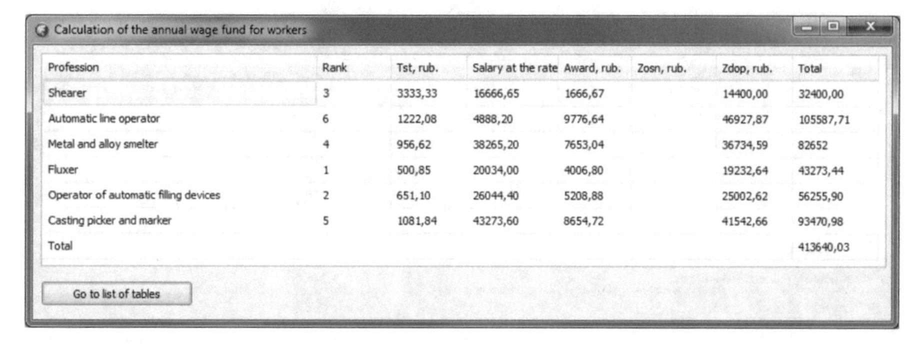

Calculation of the annual wage fund for workers

Profession	Rank	Tst, rub.	Salary at the rate	Award, rub.	Zosn, rub.	Zdop, rub.	Total
Shearer	3	3333,33	16666,65	1666,67		14400,00	32400,00
Automatic line operator	6	1222,08	4888,20	9776,64		46927,87	105587,71
Metal and alloy smelter	4	956,62	38265,20	7653,04		36734,59	82652
Fluxer	1	500,85	20034,00	4006,80		19232,64	43273,44
Operator of automatic filling devices	2	651,10	26044,40	5208,88		25002,62	56255,90
Casting picker and marker	5	1081,84	43273,60	8654,72		41542,66	93470,98
Total							413640,03

Go to list of tables

Fig. 7. Results of calculating the annual wage fund and calculating production costs.

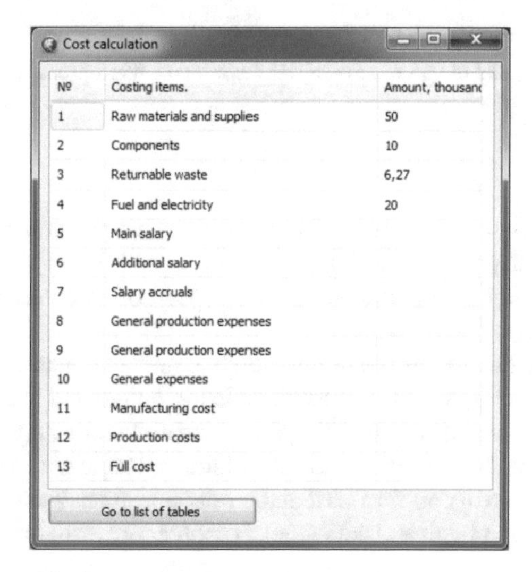

Cost calculation

№	Costing items.	Amount, thousanc
1	Raw materials and supplies	50
2	Components	10
3	Returnable waste	6,27
4	Fuel and electricity	20
5	Main salary	
6	Additional salary	
7	Salary accruals	
8	General production expenses	
9	General production expenses	
10	General expenses	
11	Manufacturing cost	
12	Production costs	
13	Full cost	

Go to list of tables

Fig. 8. Results of calculating production costs.

Thus, the developed application for automated foundry scheduling provides the following functions:

- planning of enterprise resources, including production equipment, human capital, and inventories;
- formation of the list of work to be performed and the working time fund, taking into account weekends, holidays and other non-working days;
- detailing the cost structure, including payroll, social tax, depreciation charges, overhead and general business expenses and other production costs;
- determination of the cost of non-ferrous metal smelting for the purpose of more competent pricing of products.

The application has a Windows-oriented interface, operating system Windows 10 and above, type of implementing computer: IBM PC not lower than “Pentium”, program volume 13.1 MB, programming language “Borland Delphi”.

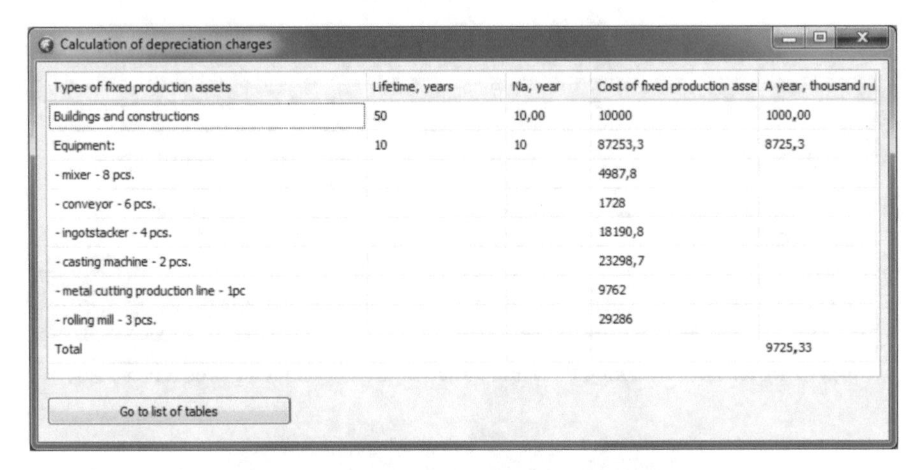

Calculation of depreciation charges

Types of fixed production assets	Lifetime, years	Na, year	Cost of fixed production asse	A year, thousand ru
Buildings and constructions	50	10,00	10000	1000,00
Equipment:	10	10	87253,3	8725,3
- mixer - 8 pcs.			4987,8	
- conveyor - 6 pcs.			1728	
- ingotstacker - 4 pcs.			18190,8	
- casting machine - 2 pcs.			23298,7	
- metal cutting production line - 1pc			9762	
- rolling mill - 3 pcs.			29286	
Total				9725,33

Go to list of tables

Fig. 9. Results of calculation of depreciation of fixed production assets.

6 Conclusion

Modern business is extremely sensitive to errors in planning and management, therefore, to make effective management decisions, especially in conditions of uncertainty and risk, it is necessary to constantly monitor all kinds of aspects of the enterprise's activities and, first of all, production.

Therefore, in the conditions of competition and market dynamism, business entities cannot refuse to create new and modernize existing automated accounting systems of industrial enterprise, capable of providing increased efficiency of production processes.

The results of the present work are the algorithm and logic of functioning of software for optimal organization of non-ferrous metal production, management and accounting of costs for materials, equipment and personnel, computer program for automated planning of foundry shop work.

A promising way of further use of the developed application can be considered its use in the form of an add-on of regular corporate information systems for the purpose of integrated management of enterprise resources.

References

1. Alchinov, A., Tavbulatova, Z., Dudareva, O., Ivanov, M.: Modern approach to enterprise information systems. Journal of Physics: Conference Series **1661**, 012164 (2020)
2. Vakhrusheva, M., Khaliev, M., Pokhomchikova, E.: Barclays' application of information system in manufacturing process. Journal of Physics: Conference Series **2032**, 012129 (2021)
3. Kravets, A., Seelantiev, A., Salnikova, N., Medintseva, I.: Redmine-based approach for automatic task distribution in the industrial automation projects. Studies in Systems, Decision and Control **418**, 261–273 (2022)
4. Aristova, N.: Intelligence in industrial automation. Automation and Remote Control **6**, 1071–1076 (2016)

5. Kovin, R., Kudinov, A., Markov, N., Miroshnichenko, E.: Information technologies in industrial enterprises production assets management. Key engineering materials **685**, 823–827 (2016)
6. Varela-Aldás, J., Chávez-Ruiz, P., Buele, J.: Automation of a lathe to increase productivity in the manufacture of stems of a metalworking company. Commu. Comp. Info. Sci. **1195**, 244–254 (2020)
7. Sinha, R., Patil, S., Vyatkin, V., Gomes, L.: A survey of static formal methods for building dependable industrial automation systems. IEEE Trans. Indus. Info. **7**, 3772–3783 (2019)
8. Krakovskaya, I., Korokoshko, J.: Assessment of the readiness of industrial enterprises for automation and digitalization of business processes. Electronics **21**, 2722 (2021)
9. Voit, N., Kirillov, S., Kanev, D.: Automation of workflow design in an industrial enterprise. Lecture Notes in Computer Science **11623**, 551–561 (2019)
10. Mejía-Neira, Á., Jabba, D., Caballero, G., Caicedo-Ortiz, J.: The influence of software engineering on industrial automation processes. Informacion Tecnologica **5**, 221–230 (2019)
11. Nezmetdinov, R., Melikov, P., Utarbaev, R.: Development of the industrial room automation system on the basis of a single computer. Lecture Notes in Electrical Engineering **857**, 92–101 (2022)

Information System for Managing Material Remuneration of Teachers and Its Analytical Potential by Example of Bauman Moscow State Technical University

Evgeniy Kostyrin(✉) and Evgeniy Sokolov

Bauman Moscow State Technical University, Building 1, 5, 2-nd Baumanskaya Street, Moscow, Russia
kostyrinev@bmstu.ru

Abstract. The algorithm developed in this article, the economic and mathematical model and tools for managing paid educational services, which coordinate the growth of teachers' salaries with the number of students in study groups, the number of study groups, the volume of classroom hours and deductions for the development of the University, allow: 1. Significantly (two times on average) to increase the funds allocated for the remuneration of teaching staff at the prestigious departments of the University and thereby solve the problem with the shortage of highly qualified teachers. 2. Dramatically increase the interest of all departments of the University in the recruitment of paying students. 3. Significantly, by more than 50%, increase the funds received for the development of the University.

Keywords: Paid Educational Services · Economic and Mathematical Model · Study Group · Teaching Staff · Teacher · Salary · Labour Stimulation

1 Introduction

Universities and educational organizations are a fundamental element of the social sphere of any state. Their development largely depends on the availability of the necessary resources, information systems for managing the material and moral remuneration of teaching staff, the methodology for making managerial decisions aimed at developing paid educational services, attracting highly qualified specialists in the subject areas of research in the field of higher education institutions and on the possibilities of organizations in the field of improvement its material and technical base. A sufficient amount of funding allows you to form a highly qualified teaching staff, regularly update and develop new training programs, use modern equipment and much more. In turn, low salaries, high moral and physical wear and tear of the main part of the material and technical base inherent in many organizations, outdated library stock and the lack of modern information and educational technologies that increase the efficiency of mastering educational material are insurmountable problems in conditions of a shortage of financial resources.

A. Gibadullin (Ed.): DITEM 2023, LNNS 942, pp. 14–24, 2024.
https://doi.org/10.1007/978-3-031-55349-3_2

Despite the high social significance of education, overcoming the funding deficit of such organizations must be solved not only by expanding state support, but also by increasing extra-budgetary funds received from paid educational activities. This circumstance determines the need for a deep and thorough study of the processes occurring in the provision of paid educational services, and the development of economic and mathematical models, as well as information systems based on them for making informed management decisions aimed at increasing the material remuneration of the teaching staff.

The problem also lies in the fact that prestigious departments at the Bauman Moscow State Technical University can recruit (there is demand) much more paid students. But due to the insufficiently high salaries of the teaching staff, there are not enough teachers to meet the increased demand for paid training.

At the same time, it should be noted that under the existing system of remuneration of teachers, the number of students in the study group is not taken into account. And the receipts of funds to the University from a group, for example, numbering 30 people, are 3 times higher than the receipts from a group of 10 people.

2 Materials and Methods

An innovative decision-making system in education that matches the growth of teachers' salaries with an increase in the volume of paid educational services provided (the number of students in study groups, classroom hours, study groups, etc.), deductions for administrative and managerial personnel and for the development of the University, has the form [1–3]:

Target Function:

$$T_{ij} = T_{bij} + \xi_{ij} \cdot \left(I_{ij} - I_{bij}\right) \rightarrow \max, \tag{1}$$

Restrictions:

$$I_{devij} = I_{devbij} + (1 - \xi) \cdot \left(I_{ij} - I_{bij}\right) \cdot \left(1 - T_{profit}\right) - \beta_{ij}, \tag{2}$$

$$\beta_{ij} = T_{ij} \cdot \varphi_{ij}, \tag{3}$$

$$\Delta C = V_{ij} \cdot \left(C_{varij} + \frac{C_{fixj}}{\sum_{k=1}^{n_j} V_{ijk}}\right) - V_{bij} \cdot \left(C_{varij} + \frac{C_{fixj}}{\sum_{k=1}^{n_j} V_{ijk}}\right), \tag{4}$$

$$I_{ij} = F_{ij} - V_{ij} \cdot \left(C_{varij} + \frac{C_{fixj}}{\sum_{k=1}^{n_j} V_{ijk}}\right), \tag{5}$$

$$F_{ij} = V_{ij} \cdot x_{ij}, \tag{6}$$

$$\Delta_{ij} = P_{bij} - \gamma \cdot x_{ij}, \tag{7}$$

$$x_{ij} = \frac{P_{bij} \cdot \left(V_{ij} - V_{bij}\right) + C_{bij} - C_{ij}}{V_{ij}}, \tag{8}$$

$$\omega_{fixij} = \frac{\frac{C_{fixj}}{\sum_{k=1}^{n_j} V_{ijk}}}{C_{varij} + \frac{C_{fixj}}{\sum_{k=1}^{n_j} V_{ijk}}}, \quad (9)$$

$$\omega_{varij} = \frac{C_{varij}}{C_{varij} + \frac{C_{fixj}}{\sum_{k=1}^{n_j} V_{ijk}}}, \quad (10)$$

$$\sum_{k=1}^{n_j} V_{ijk} \le Norm_{ij}, \quad (11)$$

$$V_{ijk} \ge 0, \quad (12)$$

$$x_{ijk} \ge 0. \quad (13)$$

In the innovative decision-making system in education (1)-(13), the following designations are used: T_{ij} is the average tariff for paid educational services for one classroom hour of work of the i-th teacher of the j-th division of an educational organization, rubles; T_{bij} is the average tariff for paid educational services for one classroom hour of work of the i-th teacher of the j-th division of an educational organization in the basic version of modelling, rubles; ξ_{ij} is the coefficient of redistribution of income growth from educational activities between the i-th teacher of the j-th division of an educational organization and the educational organization development fund; I_{ij} is income of the educational organization from the provision of paid educational services by the i-th teacher of the j-th division of the educational organization, rubles; I_{bij} is income of the educational organization from the provision of paid educational services by the i-th teacher of the j-th division in the basic version of modelling, rubles; I_{devij} is the amount of deductions directed to the development of an educational organization when providing paid educational services by the i-th teacher of the j-th division of the educational organization, rubles; I_{devbij} is the amount of deductions directed to the development of an educational organization when providing paid educational services by the i-th teacher of j-th division of an educational organization in the basic version of modelling, rubles; H is the tax rate, %; β_{ij} is the amount of deductions to extra-budgetary funds from the salary of the i-th teacher of the j–th division of the educational organization, rubles; φ_{ij} is the rate of deductions to extra-budgetary funds from the salary of the i-th teacher of the j-th division of the educational organization, the share of units; ΔC is change in the cost of paid educational services due to changes in the volume of their provision, rubles; V_{ij} is the volume of paid educational services rendered by the i-th teacher of the j-th division of the educational organization, units.; C_{varij} is conditionally variable costs of paid educational services provided by the i-th teacher of the j-th division of the educational organization, rubles; C_{fixj} is conditionally fixed costs of the j-th division of the educational organization, rubles; $\sum_{k=1}^{n_j} V_{ijk}$ is the total volume of paid educational services rendered by the i-th teacher of the j-th division of the educational organization, units; n_j is the number of varieties of paid educational services of the j-th division of the educational organization; V_{bij} is the volume of paid educational services rendered by the i-th teacher of the j-th division of the educational organization in the basic version of modelling, units; F_{ij} is the receipt of funds for paid tuition in the educational organization of students of the study group in which the i-th teacher of the j-th division of the educational organization conducts classroom classes, rubles; x_{ij} is the average annual cost of paid tuition in an educational organization for students of a study group in which the i-th teacher of the j-th division of an educational organization conducts classroom classes and which, with an increased volume of educational services of the j-th division of an educational organization, provides the same flow of financial resources as with the basic volume of these educational services, rubles; Δ_{ij} is the amount of reduction in the annual cost of paid tuition in the educational organization of students of the study group in which the i-th teacher of the j-th division of the educational organization conducts classroom classes, rubles; C_{bij} is the annual cost of paid tuition in the educational organization of students of the study group in which the i-th teacher of the j-th division of the educational organization conducts classroom classes, rubles; γ is coefficient of redistribution of discounts on paid educational services between students of paid educational courses and an educational organization; C_{bij} is the average cost of paid educational services rendered by the i-th teacher of the j-th division of the educational organization in the basic version of modelling, rubles; C_{ij} is the actual cost of paid educational services rendered by the i-th teacher of the j-th division of the educational organization, rubles; ω_{fixij} is the share of conditionally fixed costs in the structure of the cost of paid services rendered educational services by the i-th teacher of the j-th division of the educational organization; ω_{varij} is the share of conditionally variable costs in the structure of the cost of paid educational services rendered by the i-th teacher of the j-th division of the educational organization; $Norm_{ij}$ is the normative volume of educational services of the i-th teacher of the j-th subdivision of an educational organization, determined by regulatory legal acts, local regulatory acts of an educational organization, the requirements of the Sanitary rules and regulations and other regulatory documents.

3 Results

The practical implementation of the innovative decision-making system in education (1)-(13) is carried out on the example of the work of a teacher providing paid educational services in one of the federal state budgetary educational institutions of higher education (University). Receipt of funds, salary and University expenses of a teacher providing paid educational services are presented in Table 1.

Column 1 of Table 1 shows the number of students in one paid group, increasing by one according to the modelling options, so that in the last modelling option (the last row of Table 1), the number of students in the paid group is 30 people. In column 2 of Table 1, the receipt of financial resources for paid University tuition per year (300,000 rubles per year) is modelled. Thus, depending on the number of students, the volume of University admissions increases proportionally from 300,000 rubles per year per student to 9,000,000 rubles. Per year, if the number of students in the paid group is 30 people. The load on one teacher per one rate according to the standard is 880 h per year, and classroom hours per year according to the curriculum, one study group accounts for an average of 1,700 h. Thus, 1,700: 880 = 1.9 teacher rates are required to conduct classroom classes in one group per year. In the structure of the teaching staff of the educational institution in question, the main share of those providing paid educational services falls on associate professors of the department with a PhD degree, whose average salary for one rate is 76,800 rubles. Thus, the average annual salary expenses of the teaching staff are equal to 1.9 rates · 76,800 rubles. (salary per one rate) · 12 = 1,751,040 rubles. According to long-term data, the number of students in the paid group at the break-even point should be 7–9 students. In our calculations, the number of the paid group is assumed to be equal to 10 students (reserve 10%). Accordingly, with such a number of students in a paid group, 300,000 rubles are received for the development of the University · 10 students – 1,751,040 rubles (average annual expenses for teachers' salaries) = 1,248,960 rubles from one paid group. Thus, with the number of students in a paid study group equal to 10 people, the percentage of income allocated to the salaries of teaching staff is 58.37%, for the development of the University is 24.12%, and the percentage of income allocated to extra–budgetary funds (social contributions) is 17.51%.

Table 1. Receipt of funds, salary and University expenses of a teacher providing paid educational services.

Number of students in a paid group	Receipt of funds for paid tuition at the University rubles per year	Salary expenses of the teaching staff, rubles	University expenses, rubles
1	2	3	4
1	300,000	1,751,040	1,248,960
2	600,000	1,751,040	1,248,960
3	900,000	1,751,040	1,248,960
4	1,200,000	1,751,040	1,248,960
5	1,500,000	1,751,040	1,248,960
6	1,800,000	1,751,040	1,248,960
7	2,100,000	1,751,040	1,248,960
8	2,400,000	1,751,040	1,248,960
9	2,700,000	1,751,040	1,248,960
10	3,000,000	1,751,040	1,248,960
11	3,300,000	1,919,040	1,380,960
12	3,600,000	2,087,040	1,512,960
13	3,900,000	2,255,040	1,644,960
....			
29	8,700,000	4,943,040	3,756,960
30	9,000,000	5,111,040	3,888,960

In fact, at present, 43%, not 58.37%, is directed to teachers for salaries from income, and 57% is directed to the development of the University. The higher income of the University is due to the fact that 20 and sometimes 30 people study in popular departments in study groups. *And, as noted above, the salaries of teaching staff do not take into account the growth in the number of students in study groups* [4–7].

In the proposed modelling example, the authors propose to divide the income from paid educational services between the teaching staff and the University in fact, i.e. 43%

by 57%. At the same time, out of 57% of the University's income, 44% remains at the University, and 13% are contributions to extra-budgetary social funds.

According to the data presented in the Electronic University information system (Electronic University, 2022), students study at the University on a paid basis 6,268 people. Thus, the total financial resources allocated to the University will amount to 6,268 people: 10 people · 1,248,960 rubles = 782,848,128 rubles per year.

The average number of classroom hours per year is determined by the curriculum of students and is 1,700 h.

Surcharge to the average rate, average tariff and the average cost of one classroom hour are shown in Table 2.

Table 2. Surcharge to the average rate, average tariff and the average cost of one classroom hour.

Number of students in a paid group	Surcharge to the average rate for one classroom hour depending on the increase in the number of students in the group, rubles	Average tariff for paid educational services per classroom hour, rubles	The average cost of one classroom hour, taking into account the number of students in the group, rubles
1	2	3	4
1	–	873.00	1,764.71
2	–	873.00	1,764.71
3	–	873.00	1,764.71
4	–	873.00	1,764.71
5	–	873.00	1,764.71
6	–	873.00	1,764.71
7	–	873.00	1,764.71
8	–	873.00	1,764.71
9	–	873.00	1,764.71
10	0.00	873.00	1,764.71
11	75.88	948.88	1,941.18
12	227.65	1,024.76	2,117.65
13	303.53	1,100.65	2,294.12
…………………………………			
29	1,441.76	2,314.76	5,117.65
30	1,517.65	2,390.65	5,294.12

The average tariff for paid educational services for one classroom hour is 76,800 rubles: 88 h of classroom lessons per month = 873 rubles per classroom hour, which is indicated in rows 1–10 of column 3 of Table 2. According to the objective function (1) of the economic and mathematical model (1)-(13) in the absence of income of an educational organization from the provision of paid educational services, as follows from the analysis of rows 1–10 of column 2 of Table 3, the average tariff for paid educational services for one classroom hour is equal to the base value, i.e. 873 rubles. However, if an educational organization exceeds the break-even point from paid educational activities (in our case, the break-even point is equal to 10 students in a paid group, see line 10 of

Tables 1, 2, 3 and 4), the average tariff for paid educational services for one classroom hour of the i-th teacher of the j-th division of the educational organization (T_{ij}) increases to 948.88 rubles in the 11th version of modelling and further increases to 2,390.65 rubles. in the 30th version of modelling, 2.74 times compared with the basic version of modelling (2,390.65 rubles: 873.00 rubles = 2.74 times). When providing profitable educational services, the average tariff for paid educational services for one classroom hour of the i-th teacher of the j-th division of the educational organization (T_{ij}) is determined by adding an allowance to the average tariff for one classroom hour, depending on the increase in the number of students in the group indicated in the corresponding row of column 2 of Table 2, to the previous value the tariff in column 3 of Table 2. For example, for the 11th row of column 3 of Table 2, the value is 948.88 rubles = 873.00 rubles (see row 10, column 3 of Table 2) + 75.88 rubles (row 11, column 2 of Table 2). Similarly for rows 12–30 of column 3 of Table 2.

Income and the increase in receipts for the growth of salaries from the provision of paid educational services are shown in Table 3.

Table 3. Income and the increase in receipts for the growth of salaries from the provision of paid educational services.

Number of students in a paid group	Income from the provision of paid educational services, rubles per year	Income from the sale of paid educational services, rubles per month	The increase in receipts for the growth of salaries of teaching staff, rubles per month
1	2	3	4
1	-2,700,000	-225,000	–
2	-2,400,000	-200,000	–
3	-2,100,000	-175,000	–
4	-1,800,000	-150,000	–
5	-1,500,000	-125,000	–
6	-1,200,000	-100,000	–
7	-900,000	-75,000	–
8	-600,000	-50,000	–
9	-300,000	-25,000	–
10	0	0	0
11	300,000	30,000	12,900
12	600,000	60,000	25,800
13	900,000	90,000	38,700
....			
29	5,700,000	570,000	245,100
30	6,000,000	600,000	258,000

The allowance for one classroom hour of teaching staff is determined by the formula (1) of the innovative decision-making system in education (1)-(13), which is the objective function of the model. The share of funds received by the educational organization, which will be directed to the allowance for one classroom hour of work of the teaching staff, is determined by the coefficient of redistribution of income growth from educational activities between the i-th teacher of the j-th division of the educational organization and the educational organization development fund (parameter ξ_{ij} in the target function). As noted above, 43% of the increase in income from the provision of paid educational services is received to stimulate the work of teaching staff, for the development of the University is 44% of income. Contributions to extra-budgetary funds are equal to 30% of salary, which is 13% of income. So, for 11th row of column 2 of Table 2 value of 75.88 rubles = 30,000 rubles (income from the sale of paid educational services per month, see row 11, column 3 of Table 3): 170 (average number of classroom hours per month) · 0.43 (percentage of income for stimulating the work of teaching staff). Similarly, for rows 12–30 of column 2 of Table 2. So, in the latest version of the simulation, the allowance for one classroom hour of teaching staff it is 1,517.65 rubles, which exceeds the basic tariff rate for paid educational services for one classroom hour by 1.74 times (1,517.65 rubles, see the last row of column 2 of Table 2: 873 rubles, see the first row of column 3 of Table 2 = 1.74 times).

Income from the provision of paid educational services (column 2 of Table 3) the differences between the receipt of funds for paid tuition are equal (column 2 of Table 1) and expenses for the provision of paid educational services (3,000,000 rubles). As follows from the analysis of the data presented in this column of Table 3, with the number of students in a paid group not exceeding 10 people, paid educational services are unprofitable. But if a teacher provides paid educational services for the number of listeners of 11 people or more, then such educational activities bring additional income to the educational organization, therefore, as shown in formula (1) of the economic and mathematical model (1)-(13), such paid educational activities should lead to an increase in the stimulation of the work of the teaching staff [8, 9]. Column 3 of Table 3 shows the income from the sale of paid educational services per month, defined as the quotient of the division of annual income from the provision of paid educational services by 10 (the number of months in the academic year) [10–14].

The increase in receipts for stimulating the work of teaching staff (column 4 of Table 3) is calculated as the product of income from the sale of paid educational services per month (rows 11–30 of column 3 of Table 3) as a percentage of income for stimulating the work of teaching staff (43%). This means that the increase in income for stimulating work per month with the number of students in a paid group of 11 people is equal to 16,800 rubles per month and is determined as follows: 30,000 rubles (income from the sale of paid educational services per month, see row 11, column 2 of Table 3) · 0.43 (percentage of income for stimulating the work of teaching staff) = 12,900 rubles. Similarly, for rows 12–30 of column 4 of Table 3.

Teacher's salary supplement, University deductions and the increase in contributions to extra-budgetary funds are presented in Table 4.

The teacher's salary supplement, depending on the increase in the number of students in the group, presented in column 2 of Table 4, is equal to the partial division of the increase in income for stimulating the work of teaching staff (column 4 of Table 3) for 1.9 teacher's rates required for conducting classroom classes in one study group. So, for row 11 of column 2 of Table 4, the value is 6,789.47 rubles = 12 900 rubles (see row

11, column 4 of Table 3): 1.9 teacher rates. Similarly, for rows 12–30 of column 2 of Table 4.

Table 4. Teacher's salary supplement, University deductions and the increase in contributions to extra-budgetary funds.

Number of students in a paid group	Teacher's salary supplement depending on the increase in the number of students in the group, rubles	University deductions from one study group, rubles per month	The increase in contributions to extra-budgetary funds at the rate of 30%, rubles
1	2	3	4
1	–	–	–
2	–	–	–
3	–	–	–
4	–	–	–
5	–	–	–
6	–	–	–
7	–	–	–
8	–	–	–
9	–	–	–
10	0.00	0.00	0.00
11	6,789.47	13,200.00	3,900.00
12	13,578.95	26,400.00	7,800.00
13	20,368.42	39,600.00	11,700.00
....			
29	129,000.00	250,800.00	74,100.00
30	135,789.47	264,000.00	78,000.00

University deductions (column 3 of Table 4) are determined by multiplying the income from the sale of paid educational services per month (rows 11–30 of column 3 of Table 3) as a percentage of income for the University. The specified percentages of income allocated to the University are 44%. So, the University's monthly deductions for the number of students in the paid group of 11 people are equal to 13,200 rubles per month and are determined as follows: 30,000 rubles (income from the sale of paid educational services per month, see row 11, column 3 of Table 3) 0.44 (percentage of University income) = 13,200 rubles. Similarly, for rows 12–30 of column 3 of Table 4.

Column 4 of Table 4 shows the increase in contributions to extra-budgetary funds, determined by multiplying income from the sale of paid educational services per month (rows 11–30 of column 3 of Table 3) as a percentage of income for contributions to extra-budgetary funds (13%). So, for the 11th row of column 4 of Table 4 value of 3,900 rubles = 30,000 rubles (income from the sale of paid educational services per month, see row 11, column 3 of Table 3) 0.13 (percentage of income for contributions to extra-budgetary funds). Similarly for rows 12–30 of column 4 of Table 4.

4 Discussion

In accordance with the simulation results presented in Tables 1–4, teachers working in study groups in which students 20 students, wages can be increased by 67,894.74 rubles, i.e. *to ensure an almost twofold increase in wages.*

At the same time, additional deductions to the University from one study group will increase by 132,000 rubles, and from all paid students studying at the University, additional deductions will increase by 132 000 rubles 6,268 paid students: 20 students in the academic group = 41,368,800 rubles per month, or 496,425,600 rubles per year, which is 63% of deductions in the basic version of the simulation corresponding to the break-even point, when 10 students study in the study group.

If 30 students study in a study group, then according to the data presented in Tables 1–4, teachers working in such groups can be increased by 135,789.47 rubles (see last row, column 2 of Table 4), i.e. *to ensure wage growth in this case is almost three times.*

At the same time, additional deductions to the University from one study group will increase by 264,000 rubles (see last row, column 3 of Table 4), so, from all paid students studying at the University, they will grow by 264,000 rubles 6,268 paid students: 30 students in the study group = 55,158,400 rubles per month, or 661,900,800 rubles per year, *which will amount to 84.55% of deductions in the basic version of the simulation* (661,900,800 rubles per year: 782,848,128 rubles per year in the basic version 100% = 84.55%).

5 Conclusions

Thus, the approach proposed in this article, linking the growth of teachers' salaries with the number of students in the study group, the number of groups, the volume of classroom hours and deductions for the development of the University will allow:

- Significantly (on average twice) increase the funds allocated for the remuneration of teaching staff at prestigious departments of the University and thereby solve the problem with the shortage of highly qualified teachers.
- Dramatically increase the interest of all departments of the University in recruiting paid students.
- Significantly, by more than 50%, increase the extra-budgetary funds received for the development of the University.

References

1. Soekamto, H.: Nikolaeva, Irina, Abbood, Abbas, Grachev, Denis, Kosov, Mikhail, Yumashev, Alexey, Kostyrin, Evgeniy, Lazareva, Natalia, Kvitkovskaja, Angelina, Nikitina, Natalya: Professional Development of Rural Teachers Based on Digital Literacy. Emerging Science Journal **6**, 1525–1540 (2022). https://doi.org/10.28991/ESJ-2022-06-06-019
2. Sokolov, E.: Kostyrin, Evgeniy: Economic and mathematical model of management of paid educational services. Econo. Manage. Probl. Solute. **12**(2), 154–162 (2022). https://doi.org/10.36871/ek.up.p.r.2022.12.02.018

3. Allcoat, D.: Hatchard, Tim, Azmat, Freeha, Stansfield, Kim, Watson, Derrick, Von Muhlenen: Adrian Education in the Digital Age: Learning Experience in Virtual and Mixed Realities. J. Educat. Comp. Res. **59**, 073563312098512 (2021). https://doi.org/10.1177/0735633120985120
4. Chmutova, I., Myronova, O.: Methods of improving staff motivation system in educational institutions. Development Management **20** (2022). https://doi.org/10.57111/devt.20(2).2022.40-50
5. Demmans, E.: Carrie, Phirangee, Krystle, Hewitt, Jim, Perfetti, Charles: Learning management system and course influences on student actions and learning experiences. Education Tech. Research Dev. **68**, 1–35 (2020). https://doi.org/10.1007/s11423-020-09821-1
6. Sikora, Y.: Usata, Olena, Mosiiuk, Oleksandr, Verbivskyi, Dmytrii, Shmeltser, Ekaterina: Approaches to the choice of tools for adaptive learning based on highlighted selection criteria. CTE Workshop Proceedings **8**, 398–410 (2021). https://doi.org/10.55056/cte.296
7. Lysokon, I.: Optimization of the activity of a higher educational institution in the conditions of crisis phenomena: socio-economic aspect. Baltic J. Legal Soc. Sci. 40–47 (2023). https://doi.org/10.30525/2592-8813-2022-4-5
8. Eseyin, E., Eseyin, C.: Teaching retirees life adjustment pattern for lifelong educational service delivery in Rivers State, Nigeria. Hungarian Educational Research Journal (2023). https://doi.org/10.1556/063.2022.00162
9. Jahnke, I., Riedel, N., Singh, K., Moore, J.: Advancing sociotechnical-pedagogical heuristics for the usability evaluation of online courses for adult learners. Online Learning 25 (2021). https://doi.org/10.24059/olj.v25i4.2439
10. Mosiiuk, O.: Sikora, Yaroslava, Usata, Olena: Usability of program interfaces for teaching 3D graphics in a school course of informatics. Info. Technol. Learn. Tools **93**, 14–28 (2023). https://doi.org/10.33407/itlt.v93i1.5098
11. Poudel, M., Roy, D.: 3D printing and technical communication in a creative factory classroom: a case study in Japan. In: ICIET 2019: Proceedings of the 2019 7th International Conference on Information and Education Technology, pp. 92–99 (2019). https://doi.org/10.1145/3323771.3323802
12. Sosnilo, A., Mayorova, E.: Intermediate results: how students assess distance learning and whether there are prospects for its application in interactive educational environments. E3S Web of Conferences, p. 371 (2023). https://doi.org/10.1051/e3sconf/202337105063
13. Vieira, E.: Silveira, Aleph, Martins, Ronei: Heuristic Evaluation on Usability of Educational Games: A Systematic Review. Informatics in Education **18**, 427–442 (2019). https://doi.org/10.15388/infedu.2019.20
14. Zhang, T., Cummings, M., Dulay, M.: An Outreach/Learning Activity for STEAM Education via the Design and 3D Printing of an Accessible Periodic Table. J. Chemic. Edu. 99 (2022). https://doi.org/10.1021/acs.jchemed.2c00186

Intra System Links Dynamics as a Cause of Development Cyclicity: The Simplest Equations

Elena V. Slavutskaya[1(✉)] and Leonid A. Slavutskii[2]

[1] Chuvash State Pedagogical University, Cheboksary 428000, Russia
lenya@slavutskii.ru
[2] Chuvash State University, Cheboksary 428015, Russia

Abstract. The description of the systems dynamics whose behavior is determined by a large number of random parameters is relevant for many areas. At the same time, the identification of system-forming factors and intra-system connections is the most important task. To identify such a system, vertical and horizontal analysis of the data describing the system is necessary. The paper uses a simplified approach to the analysis of multidimensional random data. At the qualitative level, the possibility of analyzing the system dynamics based on changes in intra-system links is clearly shown. The system dynamics is described by the Ornstein-Uhlenbeck equation, which allows to obtain analytical solutions. The results of the theoretical analysis at the qualitative level are confirmed by the author's data obtained during the analysis of psycho diagnostic data of personal development during the crisis period. The approach allows one to visually assess trends in the system development, interpret the possibility of crisis phenomena in the system, show the cyclical nature of development, based on the inhomogeneity of changes in intra-system connections. The results obtained are a confirmation and demonstration of the universal principle of system integration and differentiation.

Keywords: System-forming Factors · Intra-system Links · System Dynamics · Development Cyclicity

1 Introduction

The system approach is widely used not only in natural science, but also in sociological, economic and psychological research. It is believed that it originates from research on human physiology [5, 30]. At the initial stage of its development, system analysis was used as a tool for studying intra-system causal (hierarchical) relationships [6, 22]. The systems theory is the basis of cybernetics [3, 8]. With the development of the mathematical apparatus, the research focus is increasingly shifting to the direction of studying the complex systems dynamics development - from the society development [4, 7] to the evolution of the Universe [16].

As a rule, the parameters (characteristics) number that determine the behavior and development of the system is very large and these parameters take random values. As

A. Gibadullin (Ed.): DITEM 2023, LNNS 942, pp. 25–36, 2024.
https://doi.org/10.1007/978-3-031-55349-3_3

a result, the systems temporal transformation (evolution) has to be described by a large number of nonlinear stochastic equations [2, 27]. The solution of such systems of equations varies greatly depending on the initial conditions and the parameters variation [1, 29]. As a consequence, the results obtained are difficult to analyze and interpret.

Therefore, a comprehensive analysis of the system composition, the identification of system-forming factors and intra-system connections is the most important task. First of all, vertical and horizontal data analysis is necessary to identify the content and structure of the system, if its structure is unknown a priori.

In the event that random parameters describing the systems behavior turn out to be related to each other linearly or monotonically, parameters grouping (clustering, splitting the system into subsystems) correlation or factor analysis can be used to reduce the number of these parameters [9, 20]. This approach is widely used in economics, sociology, psychology, and so on. Factor analysis, which was originally used in psychology [9], allows to significantly reduce the number of random signs describing the system. At the same time, independent factors are identified approximately [12], but this makes it much easier to interpret the dynamic processes occurring in the system or determining the system behavior, its evolution.

As a result of such procedures, linearization of intra-system links occurs during system identification. But this does not change the ability to evaluate nonlinear processes [21, 27] occurring in the system over finite time intervals.

In this paper, a simplified approach is applied to the system analysis of multidimensional random data based on factor analysis and the simplest dynamic equations. An attempt is made to demonstrate at a qualitative level the possibilities of systems dynamics analyzing based on changes in intra-system connections.

2 Materials and Methods

2.1 Background

System analysis of data when identifying a system (if its structure is not determined a priori) is very difficult, even if it does not imply a dynamic processes assessment. It is possible to carry out a hierarchical analysis of system parameters, but it is not always possible to identify cause-and-effect relationships. The analysis of the systems dynamics implies the identification of such connections. That is, in order to system dynamics modeling, a preliminary evaluation of the main cause-and-effect intra-system links is necessary.

In this paper, a systematic approach to the phenomena study in a general context is not considered. We are talking about an inductive research method, the processing of multidimensional random data that determine the characteristics and dynamics of the system. In this case, system analysis implies a set and algorithm of sequential operations in data processing. A set of such characteristics obtained empirically as a result of determining a certain random parameters sample acts as a system, and its structure, elements, connections between them and changes in these connections and relationships are the object of system analysis.

To assess the systematic approach possibilities to data processing (system characteristics of different levels), consider the scheme shown in Fig. 1. Suppose there is a set of data obtained by different methods, using different techniques, possibly as a result of cross-sections, longitudinal studies, etc. The scheme is presented without taking into account hypotheses and a priori information about which data belong to each level of the system, it describes the sequence of their processing and system analysis.

Most often, data is classified and separated based on a priori assumptions and attitudes. If we are talking about horizontal, cross-sections or longitudinal studies, then the time and conditions for obtaining certain parameters are necessarily taken into account. It should be borne in mind that such a division and classification of the source data is carried out a priori and is justified by the tasks set. And if we follow the inductive approach from the particular to the general, this division should be made as a result of data analysis and processing. To do this, it is necessary to highlight the most significant characteristics of the system, according to which it can be divided into elements.

In general, the parameters division characterizing the elements of the system by their type must be justified. Such justification in itself requires preliminary data set processing. To explain this thesis, we can recall sociological studies when the survey is conducted anonymously, and the survey results are divided according to the content of the answers, and then the percentage of respondents of different ages and different genders who gave one or another answer to the questionnaire questions is allocated. Thus, the data classification can be carried out even at the very initial stage on the basis of their analysis and mathematical processing.

The most important task of data preprocessing is to estimate the sample sizes, data statistical distributions, the values scales and the dimension of the features used. Based on such an assessment, mathematical tools can then be selected to obtain adequate information, that is, to draw the most reliable conclusions. The selected mathematical tools can be used comprehensively. It depends on the tasks of system analysis.

Relationships between indicators or intra-system links in the simplest case can be evaluated using correlation analysis, factor analysis based on correlation coefficients is often used. In this case, Spearman's rank correlation coefficient (nonparametric) and Pearson's correlation coefficient (parametric) can be used. This allows one to define only monotonic or linear connections. Regression analysis also most often describes linear relationships between indicators. Grouping (clustering) of signs occurs according to the same laws and the criteria.

The disadvantages of these traditional statistical methods include the following:

- Based on statistical estimates, it is difficult to assess the connections between indicators that have limited or very different numerical scales. Conclusions based on the factor analysis results are valid only when monotonic or linear relationships are evaluated.
- It is very difficult to evaluate latent (indirect) connections, their hierarchy, that is, cause-and-effect patterns, using statistical methods. Cluster analysis allows one to create a hierarchical model of parameters but this is not a level classification, it is just a grouping of indicators according to the degree of their relationship, as well as in factor analysis. This grouping is also performed according to a predetermined connections' measure.

If the connections are indirect or nonlinear, machine learning methods can be used to create a hierarchical (level) model of system characteristics. It is generally believed that

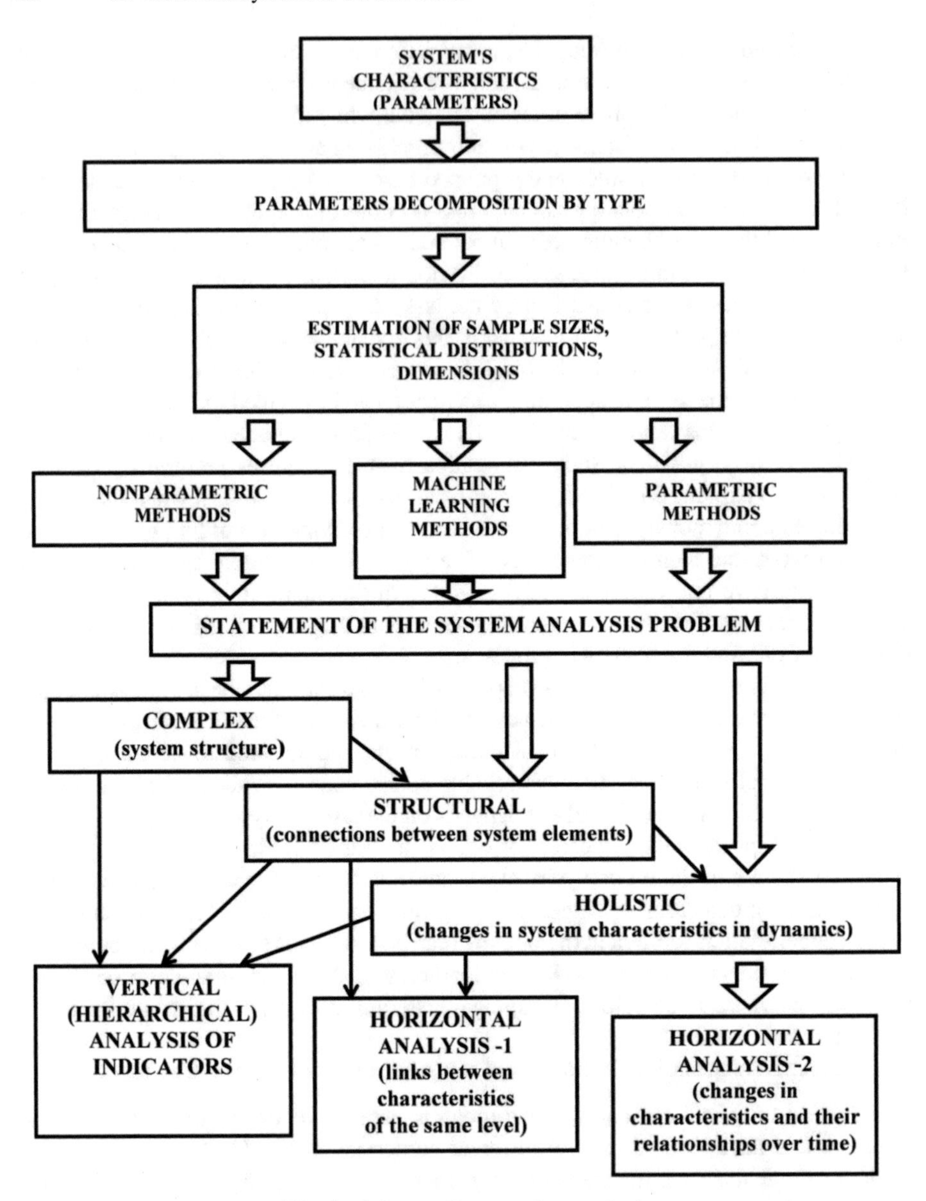

Fig. 1. Scheme of system data analysis

machine learning methods, as part of Data mining, can be used to analyze large samples ("Big Data"). Currently, the relevant tools are being developed and unified, used for sociological research. The authors have shown [24, 25] that machine learning methods and artificial intelligence elements under certain conditions can be successfully used both separately and together with traditional statistical methods for data processing and system analysis, even with small sample sizes.

If we consider a set of random data as a system, then in accordance with the classification of system analysis tasks [11] (see Fig. 1), the initial procedure is a "complex" analysis, during which the structure of the system is characterized and studied. In a narrow context, it is the data set hierarchical analysis that classifies features by levels or signs groups of the same level or type. The connections between the systems elements are not considered in the framework of the «complex» analysis. This, the initial stage of the actual system analysis by researchers in sociology or psychology, for example, is usually carried out already when setting the diagnostic task: the test results are immediately attributed to the appropriate level – individual psychological, personal, social characteristics, etc. Such an analysis is carried out a priori, without data processing. In some cases, this may not be entirely correct. An example is the personality traits in the widely used R.B. Cattell questionnaires [9]. Such a trait as excitability can be attributed to individual psychological characteristics, and the degree of group dependence can be attributed to psychosocial. Thus, in general, even within the framework of a single test, a hierarchical complex analysis of data is required.

It is considered [11] that "structural" analysis is used to study the connections between the system elements (see Fig. 1). That is, the structural model is a more advanced system analysis tool. Complex analysis can be considered as an initial stage or the basis for structural analysis. Since structural analysis involves the study of the relationships between the elements of the system, then, in addition to vertical (hierarchical) analysis, it also includes horizontal analysis corresponding to the links study between the features of one level or the division of levels into sublevels. That is, the "horizontal analysis 1" shown in the figure implies a "static" analysis of the data structure within a single control (measurement) cross section.

Often, horizontal analysis means temporary changes in indicators. But the study of dynamics is classified in system analysis as a "holistic" analysis, as opposed to a structural one. Therefore, in Fig. 1, the analysis implying a change of the system characteristics in dynamics (temporary changes) is designated as "horizontal analysis 2". Thus, longitudinal studies, for example, in economics, sociology, psychology, should, apparently, be attributed to a holistic system analysis.

2.2 Mathematical Instruments

In accordance with the general principles of systems theory, the dynamics of various systems (natural, social, economic) can, depending on specific tasks, be described by the same equations. In this paper, one of the differential equations describing dynamic processes is used in a simplified form to qualitative interpret and visual demonstrate the general universal principle of systems development - the principle of integration and differentiation [1, 10].

Such an equation can be the Ornstein–Uhlenbeck equation [28], which describes the wandering of a heavy Brownian particle taking into account friction. This is the simplest stochastic model combining the properties of the Gaussian process and the Markov process. That is, we are talking about a stationary process tending to an equilibrium state. The equation allows analytical description, interpretation and describes many processes in economics, sociology, psychology.

Without limiting generality, we will further consider the dynamic process on a psychological example. In a number of works on psychology [15, 26], the Ornstein-Uhlenbeck equation is used in the following form to assess and describe personal development within the framework of structural theory:

$$dQ(t) = k(g - Q(t))dt + \eta(t). \tag{1}$$

Here, the numerical indicators of each personality trait Q tend to an equilibrium value g over time t. $\eta(t)$ - random supplement describing independent variations in the results of psycho diagnostics. The change of function $dQ(t)$ depends on how much the indicator differs from the equilibrium value. Such an equation is written for each personality trait and each respondent. Since the process of psycho diagnostics cannot be continuous in time, the authors proceed to a discrete description of the dynamic process and numerically solve a large system of such equations. This approach allows one to assess the dynamics of personality development within the framework of the "Big Five Inventory" model, as a result, the authors identify general trends in a sample of a large number of respondents [19].

A qualitative dynamics description of changes in personality traits and an adequate interpretation can be obtained without solving a large system of equations and without taking into account stochastic additive $\eta(t)$.

The differential Eq. (1) for $\eta(t) = 0$(there is no random component, the process is regular) has the following simple analytical solution:

$$Q(t) = g \pm C\exp(-kt). \tag{2}$$

The signs $\pm$ are determined from the initial conditions $g - Q < 0, g - Q > 0$, - from above or from below Q strive for an equilibrium value. The constant C is also determined from the initial conditions at $t = 0$. The exponent in solution (2) is determined by the time constant $k = \frac{1}{T}$. It should be noted right away that the characteristic time T, that determines the dynamics of the process, and the equilibrium value g will be different for different personality traits of each respondent.

Thus, the parameters of the differential Eq. (1) vary randomly, and estimates of the averaged indicators are required for the psychological interpretation of the numerical calculations results [15].

The structural theory of personality, based on the Big Five factor model [17, 18], has been intensively implemented in recent years [14, 23]. The questionnaires of this model, especially the shortened ones, are rather a sociological tool. Therefore, to analyze the dynamics of personal development and visual interpretation of the study results, the authors of this work used Cattell's structural theory, which has been proven for many decades [9]. The personality traits that are considered are also obtained as a result of multiple factorization, but there are significantly more of them (12–16). It should be borne in mind that reducing the number of traits in factor models automatically means that these traits cannot be completely statistically independent. To study the dynamics of the relationship between personality traits, the Cattell's factor model, according to the authors, is more acceptable. It will be shown below that the number of significant factors in this model is also 4–6 [13], however, intra-system connections in the dynamics of personality development change significantly.

In factor analysis of random data, the value of each factor is determined by the following expression:

$$Z_j = a_{1j}Q_1 + a_{2j}Q_2 + \ldots + a_{rj}Q_r, \quad (3)$$

where Q_i are the numerical values of individual psychological characteristics, a_{ij} - weight coefficients that describe the relationship of the factor Z_j with these psychological characteristics. Each factor is thus a linear combination (sum) of the initial features. The coefficients a_{ij} are the factor loadings and approximately correspond to the correlation coefficients between Z_j and Q_i.

If the time dynamics of changes in each indicator Q_i is described by solution (2), then using Eq. (3), the dynamics of changes in factors $Z_j(t)$ can be obtained.

Figure 2 shows an illustration how the sum (linear combination) of two features $Z = Q_1 + Q_2$ can change over time if the values Q_1, Q_2 correspond to the solution (2), but there is heterochrony. That is, dependencies 1 and 2 have different characteristic time constants T_1, T_2, and tend to an equilibrium value g "from above" and "from below". From the example in Fig. 2, it follows that at $T_1 \neq T_2$, the time dependence (curve 3) becomes nonmonotonic. The dynamics of the values of the factor (temporary behavior in the process of personal development) acquires characteristic maxima (or minima). If a factor is determined by a linear combination of not two, but more features, then such maxima and minima may alternate in $Z(t)$. The dependence takes the form of "fluctuations" around the equilibrium value with a variable period. At the same time, the equilibrium value itself is determined by the sum of the corresponding values g_i for each psychological trait Q_i and may change over time.

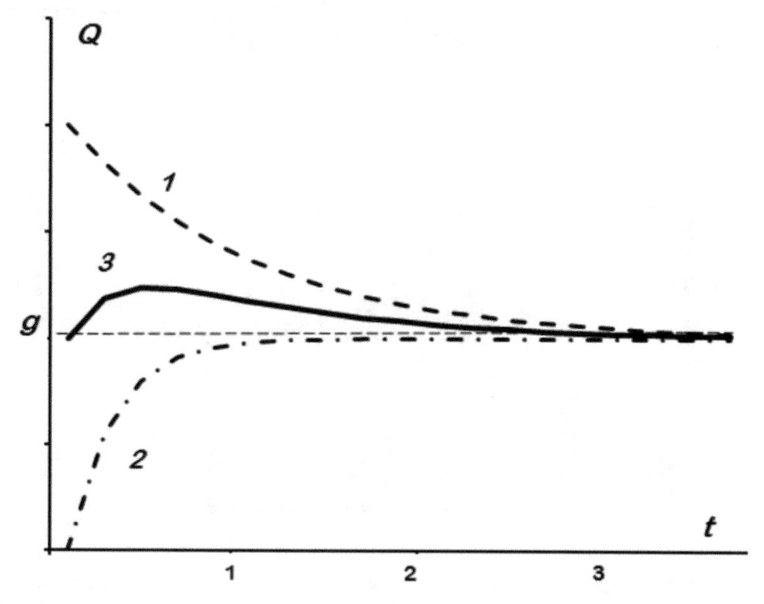

Fig. 2. The heterochrony manifestation in a linear combination of personality traits.

The degree of “severity” (amplitude) of the maxima and minima in $Z(t)$ is determined by factor loadings and heterochrony, that is, differences in characteristic times T_i for different psychological signs. The magnitude of factor loadings and the number of significant terms in the linear combination (3) determine, in general, the level of interrelationships of psychological traits or personality traits within the framework of structural theory. A high level of interrelationships of personality traits corresponds to a high level of personality structures integration, and a low level and relative statistical independence of psychological traits corresponds to their differentiation. This applies not only to the features Q_i in each factor (3), but also to the values and number of the most significant factors Z_j.

3 Results

The above theoretical analysis can be qualitatively confirmed by authors experimental data obtained as a result of a longitudinal study conducted with secondary school students for several years [13]. Figure 3 shows the factor analysis of personality traits results for students from 3rd to 8th grade of secondary school. R.B. Cattell’s 12- and 14-factor questionnaires were used depending on the age of the students. The figure shows the dynamics of the summary contribution $S_{\sum}$ of the first 4 most significant factors to the total variance. The number of these factors was selected according to the Kaiser criterion [12]. As can be seen from the figure, the temporal dynamics of the $S_{\sum}$ corresponds to the features shown above for the dynamics of factors $Z(t)$ (see Fig. 2). The dependence has a characteristic maximum at the peak of the pre-adolescent crisis [13], when personal traits, especially emotional and communicative, are maximally interconnected (integrated) (Fig. 3). After the crisis, the differentiation of personal traits in the spheres: communicative, emotional, strong-willed and intellectual is revealed.

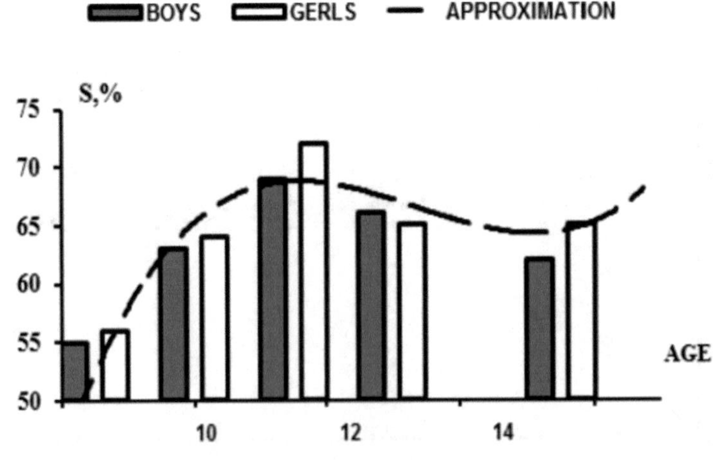

Fig. 3. The summary contribution of the 4 most significant factors in the structure of personality traits in dynamics (longitudinal study).

Moreover, Fig. 3 allows us to see and conclude about gender heterochrony of the schoolchildren mental development in the dynamics of grades 3–8. Psycho diagnostic cross-sections were carried out for boys and girls at the same time, therefore, a more pronounced peak of personality traits integration in girls during a crisis corresponds for them to greater dynamics (depth and speed of changes) in the personality structuring.

4 Discussion

The Ornstein–Uhlenbeck equation describes a random process in continuous time. When studying social processes and systems, we have to focus on discrete dimensions. Continuous monitoring is not possible, so the equation can be rewritten in finite transformations. At the same time, it should be understood that the equilibrium value can change smoothly. If these changes are much slower than the changes in the parameters that describe the variables in the equations, then the equation can be used even for non-equilibrium processes. The process becomes quasi-stationary. The dynamics of personality development discussed above based on the transformation of the Cattellian factor model in psychology is the particular example. But the approach used can be applied to social and economic processes.

During transitional and crisis periods of development, there is a fairly rapid restructuring of intra-system connections. When using psychological tools of the personality structural theory, the need to take these changes into account is clearly shown.

The basis for the analysis the processes of change, restructuring and the formation of new cross-functional links in the dynamics of human psychosocial development was modern research in the differentiation and integration direction [10]. The reason for these processes may be a violation of stability, disequilibrium of the system in age-related crisis and transition periods, which affects the structure of cross-functional relationships. The nonequilibrium state leads to the appearance of various stabilizing combinations [13]. Changes in intra– and inter-functional connections between cognitive functions, socio-psychological, individual-typological characteristics are proposed to be considered as an indicator of development. The processes of formation and structuring of intra– and inter-functional links in the age dynamics represent the universal principle of systems development - the principle of integration-differentiation.

Even taking into account the statistical, random nature of the process, which is described by Eq. (1), the model remains in equilibrium. The resulting solution for an individual respondent turns out to be random. For analysis, such a solution requires averaging over a sample of respondents. And most importantly, when solving the corresponding system of equations, it is difficult to describe the connections between personality traits. And such connections inevitably take place and have a decisive influence on the process of personal development. Each personality trait in factor models is essentially a synthesis of a whole set of psychological characteristics. The connections between them can be non-monotonic and nonlinear [24, 25]. From a mathematical point of view, from the system analysis point of view, personality traits and factors cannot be statistically completely independent. There is a heterochrony in their change in the process of human mental development. Based on the simplest analytical solution (2), the paper clearly shows how this heterochrony and changes in intra- and inter-functional connections determine the manifestation of the principle of system integration-differentiation.

5 Conclusion

The approach presented in this paper does not pretend to be an accurate quantitative description of the systems dynamics. However, it allows one to visually assess trends in the development and dynamics of the system and interpret the possibility of crisis phenomena in the system, to show the system cyclical development on the basis of uneven changes in intra-system links. Based on experimental data, estimates of the coefficients in the simplest equations used can be easily made. That is, the results obtained on the basis of such a qualitative analysis can be used as initial conditions for more accurate identification of system parameters and system-forming factors.

Such estimates can be made at separate time intervals, and by changing these coefficients, if the system is not in equilibrium (and this is necessarily present) more global temporal features of the system development can be evaluated. In addition, the results obtained are a clear confirmation and demonstration of the universal system integration-differentiation principle.

In this paper, as an example, the local psychosocial system is considered in the form of a complex of personal schoolchildren characteristics in a transitional (crisis) period. Such a system cannot be closed, and therefore equilibrium. Evaluation of this system parameters and their evolution allows us to assess at a qualitative level how the system interacts with the external environment.

References

1. Alam, M., Zada, A., Begum, S., et al.: Analysis of fractional integro-differential system with impulses. Int. J. Appl. Comput. Math. **9**, 93 (2023). https://doi.org/10.1007/s40819-023-01584-6
2. Andronov, A.A., Leontovich, E.A., Gordon, I.I., Maier, A.G.: Qualitative theory of the second order dynamic systems. John Wiley, New York (1973)
3. Ashby, W.R.: An Introduction to Cybernetics. Chapman & Hall, London (1956)
4. Beer, S.: Management science. Albus Books, London (1967)
5. Bertalanffy, L.V.: General System Theory. Foundations, Development, Applications. George Braziller, New York (1968)
6. Boulding, K.: General systems theory - the skeleton of science. General Systems **1**, 11–17 (1956)
7. Brodsky, Y.: On mathematical modeling in the humanities. In: Power Violence and Justice: Reflections Responses and Responsibilities. View from Russia: Collected Papers XIX ISA World Congress of Sociology, pp. 46–64 (2018)
8. Brodsky, Y.: From complex systems simulation to the geometric theory of behavior. In: 2020 International Conference on Mathematics and Computers in Science and Engineering (MACISE), pp. 125–132. IEEE, Madrid (2020). https://doi.org/10.1109/MACISE49704.2020.00028
9. Cattell, R.: Advanced in Cattelian Personality Theory. Handbook of Personality. Theory and Research. The Guilford Press, New York (1990)
10. Chuprikova, N.: Theory of Development: Differentiated-Integration Paradigm. Psikhologicheskiye issledovaniya **2**(3) (2009)
11. Ganzen, V.: System descriptions in psychology. Leningrad Publishing House, St. Petersburg (1984)

12. Kaiser, H.F.: The application of electronic computers to factor analysis. Educ. Psychol. Measur. **20**, 141–151 (1960)
13. Kolishev, N., Slavutskaya, E., Slavutskii, L.: Dynamics of structuring personality traits of students in the transition to the main educational school. Integration of education **23**, 3(96), 390–403 (2019)
14. Krampen, D.: The german-language short form of the big five inventory for children and adolescents - other-rating version (BFI-K KJ-F). Eur. J. Psychol. Assess. **37**(2), 109–117 (2020)
15. Kuppens, P., Oravecz, Z., Tuerlinckx, F.: Feelings change: accounting for individual differences in the temporal dynamics of affect. J. Pers. Soc. Psychol. **99**, 1042–1060 (2010). https://doi.org/10.1037/a0020962
16. Lammers, C., Hadden, S., Norman, M.: Intra-system uniformity: a natural outcome of dynamical sculpting. Monthly Notices of the Royal Astronomical Society: Letters **525**, 66–71 (2023). https://doi.org/10.1093/mnrasl/slad092
17. McCrae, R., Costa, P.: Understanding persons: From Stern's personalistics to Five-Factor Theory. Personality Individ. Differ. **169**, 109816 (2021)
18. Mõttus, R., Wood, D., Condon, D., Zimmermann, J.: Descriptive, predictive and explanatory personality research: different goals, different approaches, but a shared need to move beyond the big few traits. Eur. J. Pers. **34**(6), 1175–1201 (2020). https://doi.org/10.1002/per.2311
19. Oravecz, Z., Tuerlinckx, F., Vandekerckhove, J.: Bayesian data analysis with the bivariate hierarchical ornstein-uhlenbeck process model. Multivar. Behav. Res. **51**(1), 106–119 (2016). https://doi.org/10.1080/00273171.2015.1110512
20. Prestes, P., Silva, T., Barroso, G.: Correlation analysis using teaching and learning analytics. Heliyon **7**(11), e08435 (2021). https://doi.org/10.1016/j.heliyon.2021.e08435
21. Priyadharsini, S.: Stability analysis of fractional nonlinear dynamical systems. Advanced Mathematical Analysis and its Applications. Imprint Chapman and Hall/CRC (2023)
22. Saaty, T.: The Analytic Hierarchy Process: Planning, Priority Setting. Resource Allocation. McGraw-Hill, New York (1980)
23. Shchebetenko, S., Kalugin, A., Mishkevich, A., Soto, C., John, O.: Measurement invariance and sex and age differences of the big five inventory-2. Evidence From the Russian Version. Assessment **27**(3), 472–486 (2020). https://doi.org/10.1177/1073191119860901
24. Slavutskaya, E., Slavutskii, L., Zakharova, A., Nikolaev, E.: Integrated use of data mining techniques for personality structure analysis. Lecture Notes in Networks and Systems 345 (2021). https://doi.org/10.1007/978-3-030-89708-6_44
25. Slavutskaya, E., Slavutskii, L., Zakharova, A., Nikolaev, E.: Neural network models for the analysis and visualization of latent dependencies: examples of psycho diagnostic data processing. In: Joint Conferences XII Communicative Strategies of the Information Society (CSIS2020) and XX Professional Culture of the Specialist of the Future (PCSF2020), pp. 61–70 (2021). https://doi.org/10.1007/978-3-030-65857-1_7
26. Sosnowska, J., Kuppens, P., De Fruyt, F., Hofmans, J.: New directions in the conceptualization and assessment of personality - a dynamic systems approach. Eur. J. Pers. **34**(6), 988–998 (2020). https://doi.org/10.1002/per.2233
27. Starosta, R., Awrejcewicz, J.: Special Issue Application of Non-Linear Dynamics **12**(21), 11006 (2022). https://doi.org/10.3390/app122111006
28. Uhlenbeck, G., Ornstein, L.: On the theory of Brownian Motion. Phys. Rev. **36**(5), 823–841 (1930). https://doi.org/10.1103/PhysRev.36.823

29. Volkova, V., Leonova, A., Romanova, E., Chernyy, Y.: Engineering as a coordinating method for the development of the organization and society. Lecture Notes in Networks and Systems **184**, 12–21 (2021)
30. Vasiljev, Y., Volkova, V., Kozlov, V.: System theory and system analysis: origins and perspectives. Polytech-press, St. Petersburg (2021). https://doi.org/10.18720/SPBPU/2/id21-45

Improving the Sustainability and Safety of the City Transport System Through the Application of Computer Modeling

Irina Sippel[1](✉) and Kirill Magdin[2]

[1] Kazan Federal University, 18 Kremlin Street, Kazan 420008, Russia
irina.sippel@yandex.ru

[2] St. Petersburg State University of Architecture and Civil Engineering, 4 2nd Krasnoarmeyskaya street, St. Petersburg 190005, Russia

Abstract. The high rates of motorization observed in recent decades are accompanied by a number of negative phenomena and processes, among which the most significant are: a decrease in traffic capacity and, as a consequence, the occurrence of traffic congestion, environmental pollution with harmful substances contained in vehicle exhaust gases, and road injuries. Radical ways to improve the traffic situation, increase the sustainability and safety of the transport system, including expensive infrastructure projects, are optimal for megacities, but are not always available for small cities with low urban budget parameters. For such cities, it is advisable to use the method of simulation modeling of transport processes. This article discusses the development of a simulation model of an emergency-hazardous section of the city's road network with subsequent verification, validation of the model and conducting virtual experiments on it. A 3D simulation model was developed using GIS technologies based on the results of field studies of the parameters of transport and pedestrian flows. The validity of the computer model was assessed using two statistical criteria and was confirmed by the consistency of the results obtained during field studies and modeling. High validity indicators of the simulation model provide a greater degree of reliability of the results when conducting virtual experiments. For the analyzed emergency area, three options for improving the road situation were considered, for all of them the indicators of throughput, environmental and road safety were calculated, and the optimal option was selected.

Keywords: Transport system · Simulation model · Virtual experiment · Agent-based modeling · Safety factor · Model validation

1 Introduction

The road transport system is the most important factor in the sustainable development of the economy, social sphere and society as a whole. The steady development of the motor transport complex and the associated increase in the level of motorization, in addition to positive ones, also has negative consequences, the most significant of which

A. Gibadullin (Ed.): DITEM 2023, LNNS 942, pp. 37–48, 2024.
https://doi.org/10.1007/978-3-031-55349-3_4

are a decrease in the capacity of the motor transport system, the occurrence of traffic congestion, deterioration in environmental quality, including emissions of pollutants from vehicle exhaust gases, vibroacoustic and thermal pollution, alienation of territories for the construction of roads, bridges, interchanges and other transport infrastructure facilities, absorption of non-renewable natural resources, waste generation and the need for their disposal. Of particular danger are road traffic accidents that lead to injuries and deaths of people (pedestrians, drivers, passengers), damage and destruction of transport infrastructure and vehicles.

One of the factors causing transport problems in cities is the existing disproportion between the development of the road network and the growth of the vehicle fleet, which is especially important for large industrial centers and megacities [1].

A promising direction for increasing the sustainability of the transport system and solving the problems of the motor transport complex is the digitalization of the transport industry, including the development and widespread implementation of intelligent transport systems, Internet of Things (IoT), robotization of production processes, and the introduction of unmanned vehicles.

In recent years, one of the current trends in the digitalization of the economy is the development and application of Digital twins technology. The term "Digital twin" was first introduced by M. Grieves, who also defined the structure of Digital twin, which includes a real object (physical model), a virtual model and connections between them [2]. That is, the basis of a Digital twin is made up of digital models of physical objects that imitate their behavior in real conditions with a high degree of reliability; any Digital twin is developed on the basis of data obtained from studying the original [3].

The effectiveness of using Digital twins at all stages of the life cycle of an object has been shown, which is especially important in high-tech and labor-intensive sectors of the economy [4, 5].

Digital twin technology has found application in the aerospace industry for the development and intelligent maintenance of aircraft [6–8], in the field of robotics, including robotic manipulators and industrial robots [9–11], in industry when creating virtual machines for cyber-physical production [12], for managing a "smart" workshop [13].

The works [14–16] analyzed the effective experience of creating Digital twins for monitoring and maintaining the performance of facilities in the oil and gas sector of the economy and the petrochemical industry.

The article by Y. Liu and his colleagues shows that the use of Digital twins in the design of buildings and structures increases management efficiency in the construction industry [17].

Digital Twin technology has been successfully used to optimize production processes in the automotive industry. Based on this technology, the Russian electric vehicle KAMA-1 was developed, and currently the largest Russian automobile plant KAMAZ is developing fuel cells running on environmentally friendly hydrogen. It is noted that there is great potential for the use of Digital Twin technology in the design, production, and operation of vehicles, that is, throughout their entire life cycle [18, 19].

Many authors note that in recent years, intelligent production using big data, artificial intelligence and various types of Digital twins has begun to dominate the global economy [4, 20, 21].

Examples of the use of Digital twins to solve problems in the motor transport complex are not well represented in the literature and relate mainly to the technology for the development and use of electric vehicles, the share of which in the traffic flow is currently small, and the prospects for their mass use in the near future are ambiguous [22–25].

The works presented in the literature are devoted primarily to the creation of simulation models of the road network of large and medium-sized cities or its individual fragments. In this case, modeling is performed at the microscopic, macroscopic, and mesoscopic levels [26, 27].

In macroscopic modeling, traffic flow is considered as a single set of vehicles and is characterized by general parameters, such as average speed, intensity, and flow density. Microscopic modeling is used for small sections of the road network; it is characterized by higher accuracy in reproducing real transport processes and infrastructure [28]. Mesoscopic modeling occupies an intermediate position; the results of using mesoscopic modeling of vehicle flows are presented in [29].

Makarova's works discuss the use of macroscopic simulation modeling to optimize bus traffic in a large city with a population of more than half a million people [30, 31].

The article [32] presents the results of using the open source simulator SUMO (Simulation of Urban Mobility) to simulate vehicle movement, create traffic jams on a road map in various scenarios and identify factors that contribute to increased travel time.

The use of microscopic simulation allows us to examine in more detail the structure, features of traffic flow, and infrastructure elements of the investigated section of the highway.

The work of A. Ramos [33] discusses the development of intelligent transport systems on urban roads in Portugal and examples of the use of traffic microsimulation based on AIMSUN software.

The article [34] presents modeling of pedestrian flows based on micro-, macro- and meso-levels using the modeling environment (software product) PedSim and the agent-based architecture ABAsim.

The simulation model is a virtual twin of the road section under study and can be considered as a Digital twin prototype containing a virtual characteristic of the physical twin. The basis of Digital twin is a simulation model that has undergone a thorough verification and validation procedure, reliably reproducing the operation of a real object and the processes on the analyzed section of the road network.

To assess the validity of transport models, different authors use the correlation coefficient, Cochran's G- criterion, Seil statistics, and the Kolmogorov-Smirnov criterion [35–37].

The purpose of this work was to develop, verify and validate a simulation model of a section of the city's road network as a prototype of a digital twin and to use the constructed model to solve problems of increasing the sustainability and safety of the motor transport complex.

2 Materials and Methods

The development of computer models and the conduct of virtual experiments were carried out in the Anylogic simulation environment using the Java programming language, and the road traffic library and pedestrian library built into the software package were used.

The creation of virtual twins (simulation models) of the considered sections of the road network was carried out on the basis of agent-based and discrete-event modeling, as well as system dynamics. In agent-based modeling, each vehicle, called an agent, is examined in detail, its parameters are determined in detail, which contributes to higher modeling accuracy and a more detailed analysis of the causes and ways to solve transport problems.

The creation of simulation models was preceded by extensive (numerous) field studies on the road section, in which a quantitative assessment of the intensity of transport and pedestrian flows was carried out using video recording. Validation and verification of the model was also carried out based on the results of field studies.

The calculation of pollutant emissions from vehicle exhaust gases was carried out in accordance with the requirements of the Russian national standard GOST R 56162–2019 for the following pollutants: carbon monoxide (II), nitrogen oxides, total hydrocarbons, soot, benzo(a)pyrene, formaldehyde, sulfur oxide (IV). The calculations took into account the structure of the traffic flow, specific mileage emissions of pollutants, and the time of vehicles crossing the analyzed section of the road network [38, 39].

The safety factor was calculated as the ratio of the maximum speed of vehicles on a section of the road to the maximum speed of entry to this section.

The results of full-scale and virtual experiments were processed using the Statistica software product. The validity of the simulation model was assessed using the Cochran criterion and the Pearson correlation coefficient.

3 Results and Discussions

The level of road safety, in particular, the likelihood of road accidents leading to injuries and deaths, depends to a large extent on the ineffectiveness of the organization of interaction between pedestrians and motor vehicles [40]. The decisive factor influencing road safety is the imperfection of infrastructure, and, as a consequence, a large number of intersection points of transport and pedestrian flows. Reducing the number of conflict points where passenger, motorized and non-motorized traffic flows intersect is the most effective way to improve the safety of the transport system.

To conduct the research, a section of the road network was selected on Naberezhnye Chelny Avenue in the city of Naberezhnye Chelny. This avenue is the city's main highway and experiences higher traffic volumes than other highways, both for vehicles and pedestrians. When analyzing statistical data, it was revealed that a large number of traffic accidents have occurred at the intersection under study over the past five years, especially involving pedestrians.

To solve transport problems in this emergency area, a simulation method was chosen using the Anylogic software product. Numerous field studies of transport and pedestrian flows were carried out, as a result of which it was revealed that the highest traffic intensity is observed in the period from 16 to 18 h. At this time, long traffic jams form on the avenue, the cause of which is an artificial road hump in front of the pedestrian crossing, which slows down traffic. An analysis of the structure of the traffic flow was carried out, which showed the predominance of cars and minibuses. As a result, a simulation model was developed - a virtual twin of the road section under consideration, corresponding to the existing state of the transport situation (see Fig. 1).

Fig. 1. 3D model of the studied section of the city road network.

This 3D model is built on the basis of field research data and GIS technologies and reflects the parameters of traffic flows.

The scheme for regulating pedestrian flow in the simulation model is shown in Fig. 2.

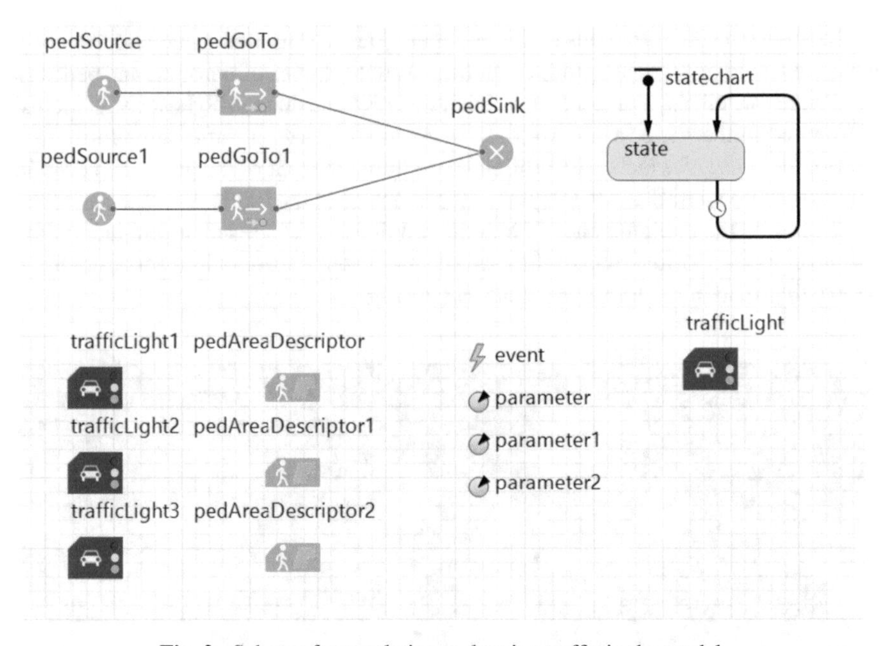

Fig. 2. Scheme for regulating pedestrian traffic in the model.

The most important stage in the development of a simulation model is its validation: the higher the validity of the model, the more reliable the simulation results are, and the higher the degree of similarity between the real object and the virtual twin. Validation, including checking the adequacy of the model of the real situation at the intersection, was carried out using two statistical parameters: the Pearson correlation coefficient and the Cochran criterion. The results are presented in Table 1 and Fig. 2.

Previously, before statistical processing, the obtained data were checked for compliance with the normal distribution law.

Table 1. Initial data and values of the Cochran criterion.

Parameter	Parameter value
Number of variances n	8
Number of results in series m	5
Number of degrees of freedom $f = m - 1$	4
Confidence probability P	0.95
Cochran criterion value:	
theoretical G_{table}	0.3910
experimental G_{exp}	0.3854

For the results obtained, the following condition is satisfied: $G_{exp} \leq G_{table}$, therefore, the dispersions under consideration are homogeneous.

Using the Statistica software, a scatterplot was constructed for the results of field studies and data obtained from modeling (see Fig. 3). The Pearson correlation coefficient was r = 0.947 with a confidence level of P = 0.95. The obtained value of the correlation coefficient confirms the homogeneity of the sample and indicates the adequacy of the model.

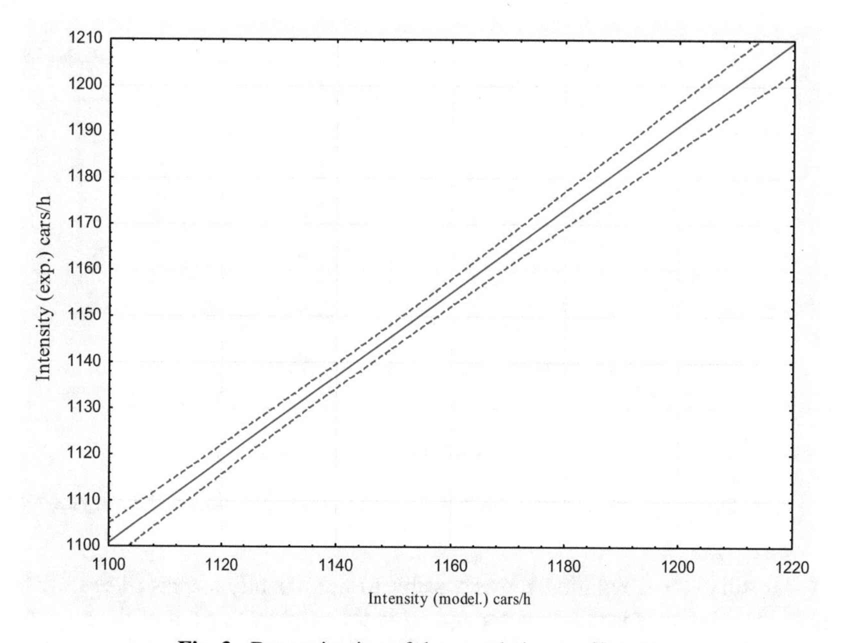

Fig. 3. Determination of the correlation coefficient.

High validity indicators of the developed simulation model indicate the possibility of its use for conducting virtual experiments and obtaining reliable results.

During simulation experiments on the model, various options for improving the transport situation on the analyzed section of the road were investigated. Three possible scenarios have been worked out in detail:

- installation of a traffic light at a pedestrian crossing;
- installation of a traffic light equipped with a pedestrian calling device;
- construction of an overhead pedestrian crossing.

When conducting virtual experiments with various options, the "State Diagram" and "System Dynamics" libraries built into the software product were used.

One of the most important advantages of a virtual twin is the visualization of the studied modeling object and the ongoing transport processes, which helps to optimize decision making.

As a result of virtual experiments, histograms were obtained for each model, illustrating in real time the parameters of traffic flow (the total number of cars crossing the road section in question, the speed of movement), as well as the total emissions of harmful substances from exhaust gases (see Fig. 4). During the simulation, the safety factor was calculated, the value of which is also recorded in the histograms. Such visualization of the results allows you to more effectively identify the advantages of the developed options and select the most optimal one. The main parameters for assessing the effectiveness of various options are: traffic intensity and average speed, mass emissions of pollutants, safety factor.

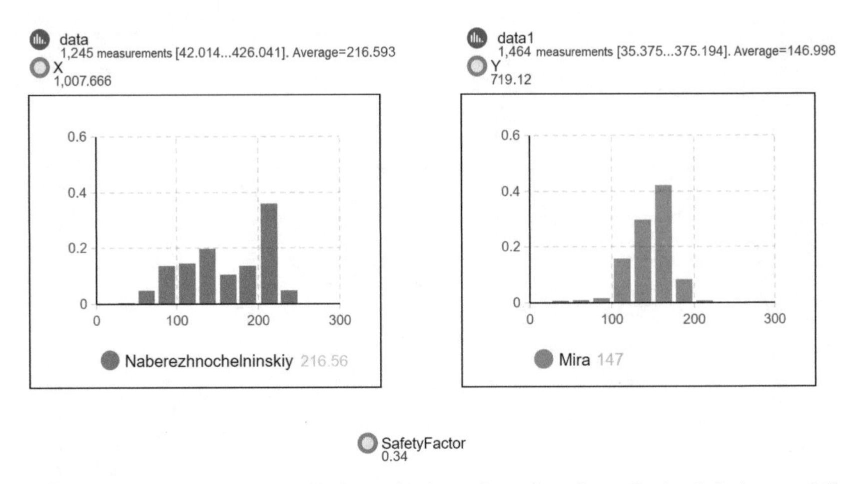

Fig. 4. Histograms of vehicle traffic intensity in various directions obtained during modeling.

As a result of the simulation, it was possible to significantly increase the safety factor (see Fig. 5). The most accident-prone option is the original version of the unregulated crossing, for which the safety coefficient is 0.34, which characterizes this section of the road as very dangerous. For all considered options, the safety factor is higher, the highest value is noted for the option with an overground pedestrian crossing: for it the safety factor value is 0.96, that is, 2.8 times higher than for the original model, which makes the road section non-hazardous.

Analysis of the results of virtual experiments (see Table 2) allows us to conclude that the unregulated transition option is also the most unecological, as it is characterized by the highest total emissions of pollutants from vehicle exhaust gases.

The most optimal option is with complete separation of traffic and pedestrian flows, which involves the construction of an overhead pedestrian crossing. This option is characterized by the highest throughput and environmental safety. In addition, when this option is implemented, road traffic accidents involving pedestrians are practically impossible, with the exception of cases involving gross violation of traffic rules by pedestrians. An obstacle to its practical implementation may be the high cost of construction. The option of installing a pedestrian calling device is inferior in terms of all the indicators to an overpass, however, its important advantage is the relatively low cost of implementation, which is important for cities with limited budget funding.

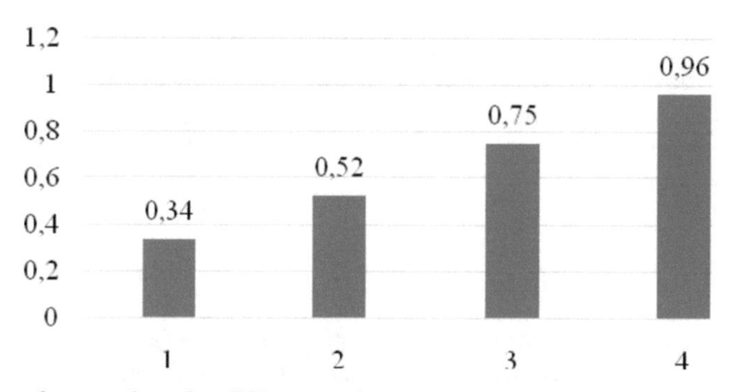

Fig. 5. Safety factor values for different options: 1 – unregulated pedestrian crossing; 2 – adjustable pedestrian crossing; 3 – pedestrian crossing with a calling device; 4 – elevated pedestrian crossing.

Table 2. Results of computer experiments.

Parameter	Unregulated pedestrian crossing	Adjustable pedestrian crossing	Pedestrian crossing with a calling device	Overhead pedestrian crossing
Total intensity, auto/h	2709	3093	3663	3903
Average travel time, min	3.6	2.7	1.3	0.9
Maximum travel time, min	7.1	5.6	2.8	1.1
Total emissions of pollutants, g	1726.2	1593.6	1121.1	750.0

Thus, at the prototyping stage, microscopic simulation models can be successfully used to solve problems of increasing the sustainability, environmental and road safety of the motor transport complex.

4 Conclusions

The use of computer modeling in the Anylogic environment is considered to solve the problem of increasing the stability and safety of the transport system with the development of a microscopic simulation model (virtual twin) of a section of the city's road network. Three options for improving the traffic situation on an emergency-hazardous section of the road are proposed; for all, the indicators of throughput, environmental and road safety are calculated. High model validity indicators ensure the reliability of the results of virtual experiments. The results obtained indicate the feasibility of using microscopic simulation models to develop solutions to improve the efficiency of managing the city's transport system.

Such simulation models can serve as the basis for the development of digital twins of sections of the city's road network, and this work is relevant within the framework of the digitalization strategy of the transport industry. Prospects for further research should include the establishment of reliable feedback between the simulation model and the real object being modeled.

References

1. Makarova, I., Magdin, K., Mavrin, V., et al.: Improving the city's transport system safety by regulating traffic and pedestrian flows. Lecture Notes in Networks and Systems **195**, 518–527 (2021)
2. Grieves, M.: Virtually Perfect: Driving Innovative and Lean Products Through Product Lifecycle Management. USA, Space Coast Press, Cocoa Beach - FL (2011)
3. Grieves, M.: Digital twin: Manufacturing excellence through virtual factory replication. White Paper, 1–7 (2014)
4. Jones, D., Snider, C., Nassehi, A. et al.: Characterising the digital twin: a systematic literature review. CIRP J. Manuf. Sci. Technol. **29**(A), 36–52 (2020)
5. Fu, Y., Zhu, G., Zhu, M., et al.: Digital twin for integration of design-manufacturing-maintenance: an overview. Chinese J. Mechan. Eng. **35**, 80 (2022)
6. Xiong, M.L., Wang, H.W., Fu, Q., et al.: Digital twin–driven aero-engine intelligent predictive maintenance. The Int. J. Adva. Manuf. Technol. **114**(l.1), 1–11 (2021)
7. Cai, H.X., Zhu, J.M., Zhang, W.: Quality deviation control for aircraft using digital twin. J. Comput. Inf. Sci. Eng. **21**(3), 031008 (2021)
8. Liao, M., Renaud, G., Bombardier, Y.: Airframe digital twin technology adaptability assessment and technology demonstration. Eng. Fract. Mech. **225**, 106793 (2019)
9. Matulis, M., Harvey, C.: A robot arm digital twin utilising reinforcement learning. Comput. Graph. **95**(1), 106–114 (2021)
10. Kaigom, E.G., Rossmann, J.: Value-driven robotic digital twins in cyber-physical applications. IEEE Trans. Industr. Inf. **17**(5), 3609–3619 (2021)
11. Wu, W., Li, M., Hu, J., et al.: Research on guidance methods of digital twin robotic arms based on user interaction experience quantification. Sensors **23**(17), 7602 (2023)
12. Cai, Y., Starly, B., Cohen, P., et al.: Sensor data and information fusion to construct digital-twins virtual machine tools for cyber-physical manufacturing. Procedia Manufacturing **10**, 1031–1042 (2017)
13. Leng, J., Zhang, H., Yan, D., et al.: Digital twin-driven manufacturing cyber-physical system for parallel controlling of smart workshop. J. Ambient. Intell. Humaniz. Comput. **10**(3), 1155–1166 (2019)
14. Wanasinghe, T.R., Wroblewski, L., Petersen, B.K., et al.: Digital Twin for the Oil and Gas Industry: Overview, Research Trends Opportunities, and Challenges. Computer Science IEEE Access (2020)
15. Gao, L., Jia, M., Liu, D.: Process digital twin and its application in petrochemical industry. J. Softw. Eng. Appl. **15**, 308–324 (2022)
16. Correia, J.B., Rodrigues, F., Santos, N., et al.: Data management in digital twins for the oil and gasindustry: beyond the OSDU data platform. J. Info. Data Manage. **13**(3), 405–420 (2022)
17. Liu, Y., Sun, Y.H., Yang, J., et al.: Digital twin-based ecogreen building design. Complexity **10**, 1–10 (2021)
18. Biesinger, F., Weyrich, M.: The facets of digital twins in production and the automotive industry. In: 23rd International Conference on Mechatronics Technology (ICMT) (2019). https://doi.org/10.1109/ICMECT.2019.8932101

19. Piromalis, D., Kantaros, A.: Digital twins in the automotive industry: the road toward physical-digital convergence. Applied System Innovation **5**(65) (2022). https://doi.org/10.3390/asi5040065
20. Qi, Q., Tao, F.: Digital twin and big data towards smart manufacturing and industry 4.0: 360 degree comparison. IEEE Access **6**, 3585–3593 (2018)
21. Winter, S., Tomko, M.: Beyond digital twins – a commentary. Environ. Planning B: Urban Analytics and City Science **46**(2), 395–399 (2019). https://doi.org/10.1177/2399808318816992
22. Shikata, H., Yamashita, T., Arai, K., et al.: Digital twin environment to integrate vehicle simulation and physical verification. SEI Technical Review **88**, 18–21 (2019)
23. Ali, W.A., Roccotelli, M., Fanti, M.P.: Digital twin in intelligent transportatio systems: a review. In: 8th International Conference on Control, Decision and Information Technologies (CoDIT) (2022). https://doi.org/10.1109/CoDIT55151.2022.9804017
24. Rudskoy, A., Ilin, I., Prokhorov, A.: Digital twins in the intelligent transport systems. Transportation Research Procedia **54**, 927–935 (2021)
25. Marai, O.E., Taleb, T., Song, J.: Roads infrastructure digital twin: a step toward smarter cities realization. IEEE Network **35**(2), 136–143 (2021)
26. Sun, L., Cheng, Z., Zhang, K: Modeling and analysis of human-machine mixed traffic flow considering the influence of the trust level toward autonomous vehicles. Simul. Model. Pract. Theory **125**, 102741 (2023)
27. Ni, D.: Traffic flow theory. Characteristics, Experimental Methods, and Numerical Techniques 412 (2016). https://doi.org/10.1016/C2015-0-01702-6
28. Al-Dabbagh, M.S.M., Al-Sherbaz, A., Turner, S.: The impact of road intersection topology on traffic congestion in urban cities. Intelligent Systems and Applications. Advances in Intelligent Systems and Computing, pp. 1196–1207. Springer, Cham (2019)
29. Varga, B., Doba, D., Tettamanti, T.: Optimizing vehicle dynamics co-simulation performance by introducing mesoscopic traffic simulation. Simul. Model. Pract. Theory **125**, 102739 (2023). https://doi.org/10.1016/j.simpat.2023.102739
30. Makarova, I., Pashkevich, A., Shubenkova, K: Ensuring sustainability of public transport system through rational management. Procedia Engineering **78**(4), 137–146 (2017)
31. Makarova, I., Shubenkova, K, Gabsalikhova, L.: Analysis of the city transport system's development strategy design principles with account of risks and specific features of spatial development. Transport Problems **12**(1), 739–750 (2017)
32. Lopez, P., Behrisch, M., Bieker-Walz, L., et al.: Microscopic traffic simulation using SUMO. In: 21st International Conference on Intelligent Transportation Systems (ITSC) (2018). https://doi.org/10.1109/ITSC.2018.8569938
33. Ramos, A.L., Ferreira, J.V., Barceló, J.: Modeling & simulation for intelligent transportation systems. Int. J. Model. Optimiz. **2**(3), 274–279 (2012)
34. Kormanovál, A., Varga, M., Adamko, N.: Hybrid model for pedestrian movement simulation. In: The 10th International Conference on Digital Technologies (2014). https://doi.org/10.1109/DT.2014.6868707
35. Novikov, I.A., Shevtsova, A.G., Kravchenko, A.A., et al.: Development of a procedure for adapting a model of adjustable intersection. The Russian Autom. Highw. Indu. J. **17**(6), 726–735 (2020). https://doi.org/10.26518/2071-7296-2020-17-6-726-735
36. Sargent, R.: Verification and validation of simulation models. Proceedings - Winter Simulation Conference **37**(2), 166–183 (2011). https://doi.org/10.1109/WSC.2010.5679166
37. Casas, P.F.: A continuous process for validation, verification, and accreditation of simulation models. Mathematics **11**(4), 845 (2023). https://doi.org/10.3390/math11040845
38. Makarova, I., Shubenkova, K, Mavrin, V., et al.: Environmental safety of city transport systems: Problems and influence of infrastructure solutions. Lecture Notes in Networks and Systems **68**, 24–34 (2019)

39. Makarova, I.V., Gabsalikhova, L.M., Sadygova, G.R., et al.: Ways to improve safety and environmental friendliness of the city's transport system. IOP Conference Series: Materials Science and Engineering **786**(1), 012072 (2020)
40. Ghanim, M.S., Abu-Eisheh, A.: The impact of mid-block crossing on urban arterial operational characteristics using multimodal microscopic simulation approach. In: 5th International Conference on Modeling, Simulation and Applied Optimization (ICMSAO), pp. 1–5 (2013)

Development of Software for the Analysis of Socio-Economic Indicators Based on Neural Networks

Yury Shvets[1], Victoria Perskaya[2], Itao Tao[3], Dzhannet Shikhalieva[4], Dmitry Morkovkin[2](✉), Tatyana Shchukina[5], and Yaroslav Zubov[2]

[1] V.A. Trapeznikov Institute of Control Sciences if Russian Academy of Science, 65 Profsoyuznaya Street, Moscow 117997, Russia

[2] Financial University Under the Government of the Russian Federation, 49/2 Leningradsky Avenue, Moscow 125167, Russia
morkovkinde@mail.ru

[3] Shenzhen University, 3688 Nanhai Ave, Shenzhen, Guangdong, China

[4] Moscow State University of Humanities and Economics, 49 Losinoostrovskaya Street, Moscow 107150, Russia

[5] Institute of State and Law of the Russian Academy of Sciences, 10 Znamenka Street, Moscow 119019, Russia

Abstract. This research highlights the necessity of implementing neural networks in all areas of decision-making. In this case, we focus on developing a neural network for analyzing the effectiveness of decision-making in socio-economic systems, using the healthcare sector as an example. To prepare for development, it is essential to initially define the target indicator for forecasting and the set of factors influencing it. We have chosen four mortality indicators to evaluate across different socio-economic groups: infant mortality, mortality from tuberculosis, mortality from cardiovascular diseases, and mortality from neoplasms, including malignant ones. After collecting the initial data, the research describes the working principle and the development sequence of the Future Analytics program and presents the results of the analysis. The program is versatile in that it can analyze any data, but they must be closely related to the studied object.

Keywords: Neural Networks · Socio-Economic Security · Management · Healthcare System · Chronic Non-Communicable Diseases

1 Introduction

The healthcare system is an integral component of the socio-economic security of any state. Given the increasing global relevance of humanity's pervasive challenges, states require a reevaluation of approaches to address them. For example, the existing burden of chronic diseases continues to escalate each year, despite significant financial investments by states in combating them. Considering the constantly changing social and economic conditions, there is a pressing need for in-depth and comprehensive study of socio-economic healthcare systems. This is essential to better understand and manage these

A. Gibadullin (Ed.): DITEM 2023, LNNS 942, pp. 49–65, 2024.
https://doi.org/10.1007/978-3-031-55349-3_5

systems, efficiently utilize their resources, improve the quality of services they provide, and ultimately enhance the health and well-being of the population.

2 Materials and Methods

To prepare the article, a comprehensive analysis of contemporary research and scientific publications on the application of artificial intelligence and neural networks in the field of socio-economic analysis was conducted. Primary attention was given to works published in peer-reviewed journals over the past five years. Open databases were also utilized to identify current trends in this field. Quantitative and qualitative analytical methods were applied, including statistical analysis and critical thinking methods.

The program described in the study was developed based on an artificial recurrent neural network of complex architecture using the Python programming language.

3 Results

Understanding the prevalence and mortality rates, as well as their nature, scale, and social impact, is crucial for assessing healthcare systems. These health indicators play a central role in discussions about socio-economic conditions, serving a vital role in evaluating the standard of living, social equilibrium, and overall societal development.

Disease prevalence provides a dynamic picture of the population's health status, necessary for identifying health problems, social damage, and determining the healthcare system's needs, including necessary medical services and preventive programs.

On the other hand, mortality rates indicate the number of deaths within a population over a specific period and serve as a powerful indicator for assessing the impact of disease prevalence and healthcare quality. Mortality indicators, which can be general or specific depending on causes, age, or demographics, provide crucial data for assessing societal health and healthcare planning.

Both disease prevalence and mortality need to be assessed and interpreted in a socio-economic context, considering factors such as income levels, education, and social environment that influence healthcare accessibility and quality, lifestyle choices, as well as the risk of disease or death from specific conditions. These factors, challenging to quantify, nonetheless play a crucial role in understanding and improving healthcare services.

The mentioned indicators are influenced by a complex set of socio-economic factors, including income levels, education, social environment, and economic stability. It is recognized that the goals of national policy in any area are partially determined by the issues and needs the country faces [1].

Let's examine the current healthcare challenges that are relevant to the Russian Federation [2].

1. Population stability amidst increasing average age.

In terms of birth rates and family planning culture, Russia is similar to Europe, the United States, and Japan. However, a lower life expectancy and a higher mortality rate place it in line with developing countries. The decline in birth rates, characteristic of Western countries, is not offset in Russia by an increase in life expectancy and an improvement in healthcare. Even with a favorable birth rate trend, according to UN projections, Russia's population growth will gradually decline by the second half of this century, reaching almost zero by the year 2100, similar to European countries. This suggests that relying on an increase in birth rates as a means to prevent depopulation in modern conditions is impractical.

2. The necessity of extending active longevity and personalized nutrition.

Given the demographic trends described above, it becomes critically important to extend the economically active period of elderly individuals. To achieve this, it is necessary to increase the expected duration of healthy life, influenced not only by income levels and employment guarantees but also by social self-realization and the quality of the surrounding environment. Factors negatively affecting the duration of a healthy life include the prolonged period of economic recovery after the 1990s crisis, persistent risk factors such as risky behaviors (alcoholism, smoking, insufficient physical activity), and the unavailability of primary prevention measures for several common diseases.

3. Acceleration of innovation cycles as a factor in increasing the need for knowledge renewal.

The intensification of global competition and the growth of consumer demand contribute to the acceleration of innovation activity. This leads to the shortening of product life cycles and the transformation of models for the development, implementation, and dissemination of innovations in all sectors of the world economy. Despite broad international and interregional connections, there is an increasing emphasis on developing proprietary innovative solutions that are independent of external economic and political shifts, ensuring economic growth and enhancing competitiveness in global markets.

The rapid adoption of new technological solutions results in the structural transformation of the labor market, changes in the concepts of "profession" and "work," and a constant demand for new skills. In this context, the concept of lifelong learning becomes relevant, giving rise to new educational formats (online and hybrid learning, microlearning, multimedia learning, etc.) and interactive educational products based on artificial intelligence and augmented/virtual reality technologies.

4. The increasing significance of cybersecurity amidst the digitization of all human activities.

Long-term trends indicate a growing focus on cybersecurity in the digital space. Russia ranks fifth in the International Telecommunication Union's Global Cybersecurity Index, earning top scores in categories such as the quality of legislative frameworks, education, information campaigns, cybersecurity potential, and partnership development. Cybersecurity solutions from Russian companies, as well as secure electronic documentation, and domestic platforms for social services, can compete globally with Western and Asian counterparts.

Similar to the global scenario, Russia faces a shortage of professionals in the field of information security. The majority of companies (80%) are in the early stages of digital transformation, indicating a growing demand for specialists in this field [3].

The challenges in healthcare are not limited to this list and can be supplemented with the previously mentioned global issues, such as high mortality rates, the prevalence of non-communicable diseases, population aging, and others.

Now let's consider the development goals of the healthcare system in Russia for 2022–2024. According to the Decree of the President of the Russian Federation dated July 21, 2020, by 2030, it is necessary to ensure a stable population growth and increase the average life expectancy to 78 years. As noted earlier, against the backdrop of an unprecedented increase in mortality in 2021, the average life expectancy was 70.1 years. In the current circumstances, in the next two years, it is crucial first to halt the rise in mortality and then actively work towards reducing it.

The experience of past years demonstrates the effectiveness of programmatic and targeted (project-based) approaches in healthcare. For example, thanks to the implementation of the priority project "Health" from 2006 to 2012, the average life expectancy increased by 3.4 years (from 65.4 to 68.8). Programs aimed at reducing infant mortality from 2012 to 2019 decreased it by 40% (from 8.2 to 4.9 cases per 1000 live births) [4].

The current national project "Healthcare" and the program "Modernization of Primary Care" play a crucial role, but their main focus is concentrated on creating and updating the infrastructure of medical institutions. The measures taken to address the staffing issue (a primary factor in healthcare accessibility) in the national project are evidently insufficient, as confirmed by the continued decline in the number of professionals in the sector: from 2018 to 2020, the overall number of doctors and nurses decreased by 18,000 people [5].

Considering the increasing demand from citizens for widely accessible free medical assistance (due to the growing number of patients and a reduction in the real income of the population), the primary directions for reducing mortality should be defined as follows:

- Ensuring the Accessibility of Primary Medical and Sanitary Care (PMSC) provided in outpatient settings, including polyclinics and medical offices, feldsher-obstetric points, as the most common form of medical assistance. To achieve this, it is imperative to promptly address the shortage of medical specialists at the primary level and establish a comprehensive system of medication provision (free distribution of prescription drugs to the entire population, not just those eligible for benefits).

- Ensuring the accessibility of emergency and urgent medical care based on emergency departments in hospitals. This will require organizing the operation of these departments in such a way that necessary diagnostics and treatment are provided amid an increasing flow of patients.
- Substantial increase in the volume and improvement of the quality of training for mid-level medical personnel. In the context of a shortage of doctors and the extension of their training periods, mid-level medical personnel can partially alleviate the workload of doctors by performing certain general labor functions.

Currently, there is no tool that would allow for a precise assessment of the effectiveness of funds invested in various federal programs. Therefore, to predict the level of the target indicator, in this study – the mortality rate, we need to conduct a comprehensive analysis of available socio-economic indicators, which can be grouped into several categories. We conducted our research based on the indicators of the city of Moscow [6–8].

1. Accessibility and quality of healthcare utilized by individuals. These data are primarily reflected in official statistics, including indicators such as the number of hospital organizations, the number of hospital beds per 10,000 people, the number of outpatient and polyclinic medical organizations, the capacity of outpatient and polyclinic medical organizations, life expectancy at birth, the number of pregnancies with abortive outcomes, the number of doctors of all specialties and mid-level medical personnel, and others.
2. Demographic indicators. These specifically reflect the delayed impact of healthcare system activities and the standard of living on the population: population size, mortality rate, the number of births per year, the number of registered deaths, and others.
3. Economic situation in the region. This category is extensive, and its content may vary, as the economy encompasses every aspect of societal life. We have selected the following indicators: availability of fixed assets at the end of the reporting year, Gross Regional Product (GRP), cash income, consumer and other incomes, the consumer price index for goods and services, the share of small and medium-sized enterprises in GRP, the poverty rate, and others.
4. Environmental situation. One of the most crucial indicators influencing public health. The "cleanliness" of the environment can be compared to preventive measures undertaken for the population – the better the ecology (the more preventive measures), the less the need to seek help at the hospital, i.e., a lower overall morbidity rate.

Indicators: emissions of pollutants, the proportion of captured and neutralized pollutants, use of fresh water, and others.

5. Technological Development. The future of medicine and the pace of progress depend on this. As mentioned earlier, innovative technologies in medicine emerge precisely due to digitization and the implementation of artificial intelligence. Therefore, continuous development in these fields is necessary, enabling a shift toward an intensive development path for the healthcare system on the way to a value-oriented approach.

Indicators of technological and scientific development: Number of Internet users, the proportion of citizens using electronic means to access state and municipal services, the coefficient of inventive activity, research and development expenses, and others.

6. Lifestyle. Also, one of the most crucial indicators affecting human health. It is known that many things depend on the lifestyle, including the manifestation of genetic predispositions to various diseases. For example, a person leading a sedentary lifestyle, consuming alcohol and highly processed foods, is more likely to experience a myocardial infarction with a family history of such incidents than someone who leads a healthy lifestyle.

From such simple things as healthy sleep, regular walks, and a clean balanced diet, a lot actually depends.

Indicators of the population's lifestyle: Meat and meat product consumption, Milk and dairy product consumption, Vegetable and melon crop consumption, the number of swimming pools, the number of children who have had a summer break, and others.

As for investments in reducing morbidity and mortality, it is noteworthy that a significant portion of funds (62%) is allocated to combating oncological diseases, which is highly relevant both for Russia and the rest of the world. Next is the development of child healthcare and digital infrastructure, but with a significant lag. The funding for other projects does not exceed 5% of the total amount [9].

It is unclear how the outcome would have changed if the distribution of funding had been slightly shifted towards the development of digital infrastructure. As known, crucial elements for transitioning to value-based healthcare include digitization and artificial intelligence. In other words, modern equipment can aid in early tumor detection, and the implementation of continuous monitoring through wearable devices for those at risk of heart attacks or strokes can save numerous lives.

To answer these questions and make more effective management decisions, an automated model is required. This model should be capable of calculating all available facts and making reliable predictions based on them for the most important and effective healthcare system indicators, such as mortality or life expectancy.

In accordance with the goal of our research, we have developed a program that models the situation occurring with the dependent variable based on the influence of numerous factors. In this case, there is an impact of a large number of factors, allowing for a more objective representation of the predicted situation. Therefore, it makes sense to build the program on neural networks.

Recurrent Neural Networks (RNNs) are a highly advanced tool in the field of artificial intelligence actively applied in natural language processing. These networks are unique in that they can "remember" and analyze sequences of data, such as sentences or phrases, and predict the next element in the sequence based on the analysis.

However, it is worth noting that there are various types of RNNs. The most popular and effective among them are Long Short-Term Memory (LSTM) networks. By nature, LSTMs are still RNNs, but they are designed in a way to better retain information about long-term dependencies in data. In other words, LSTMs can "remember" more information from the past, making them particularly valuable for complex sequences. We use this type of recurrent network, considering the memory issue mentioned above [10].

The outcome of this research is the FutureAnalytics (FA) program, built on recurrent neural networks. Its idea is to process a large dataset for analytics and forecasting. Currently, there is no software that could simultaneously consider all parameters affecting the quality of life, providing a forecast for further development.

To forecast mortality indicators, a complex architecture artificial neural network was designed. Let's begin describing the structure of the neural network with an explanation of the input neurons; in the utilized neural network, there were 5 inputs. Initially, the data was divided into semantically meaningful groups of factors. Subsequently, each of these factor groups was used as a separate dataset for an individual input layer (a layer is a set of neurons). Activation functions are not used in the neurons of the input layer for this task. We transpose all the data; expenditures on federal programs are considered cumulative, and missing data is treated as equal to the data from the previous period [11].

Let's exclude columns that are not influencing factors and identify target variables. Next, all values, except for the target variables, undergo standardization, which may result in a change in data dimensionality.

In this case, we have identified 4 target variables:

- Infant mortality (multiply the indicator by 100 to standardize);
- Tuberculosis mortality (multiply the indicator by 100 to standardize);
- Mortality from circulatory diseases;
- Mortality from neoplasms, including malignant ones.

Then, the data from the input layers go into layers that are individual small models of recurrent networks (each layer of LSTM represents a small neural network). Recurrent neural networks are a type of neural network where connections between elements form a directed sequence. This allows processing series of events over time or sequential spatial chains. Recurrent networks can use their internal memory to handle sequences of arbitrary length. In this task, long short-term memory (LSTM) neural networks are used.

The architecture of our model is reflected in Fig. 1.

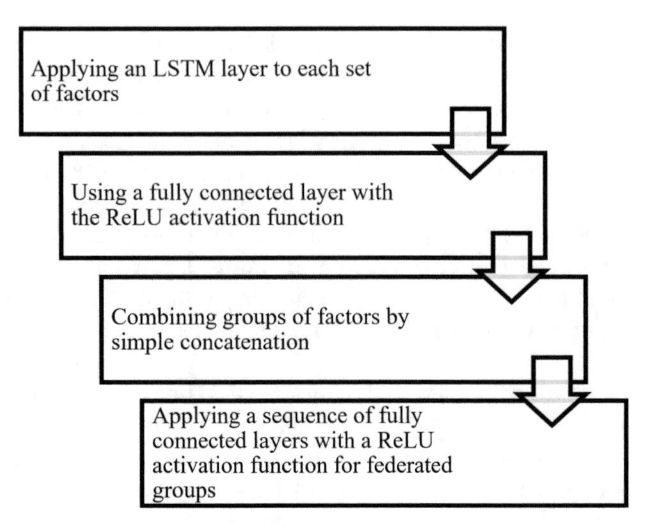

Fig. 1. Architecture of the LSTM network for the FutureAnalytics program.

Let's take a closer look at these stages. In LSTM, there are three filters that allow protecting and controlling the state of the cell. The first step in LSTM is to determine what information should be discarded from the cell state. This decision is made by the sigmoid layer called the "forget gate layer." It looks at h_{t-1} and x_t and returns a number from 0 to 1 for each number in the cell state C_{t-1}. 1 means "fully preserve," and 0 means "fully discard."

The next step is to decide what new information will be stored in the cell state. This step consists of two parts. First, the sigmoid layer called the "input layer gate" determines which values should be updated. Then, the tanh layer constructs a vector of new candidate values.

$\tilde{C}_t$, which can be added to the cell state. It's time to replace the old cell state C_{t-1} with the new state C_t. We multiply the old state by f_t, forgetting what needs to be forgotten. Then, we add $i_t * \tilde{C}_t$. These are the new candidate values multiplied by t—how much we want to update each of the state values.

Finally, we need to decide what information to output. The output will be based on our cell state, and some filters will be applied to them. First, we apply a sigmoid layer, which determines what information from the cell state we will output. Then, the cell state values go through a tanh layer to output values ranging from -1 to 1, and they are multiplied by the output values of the sigmoid layer, allowing only the required information to be output.

Each set of factors goes through its LSTM layer so that the network can thoroughly study and consider patterns related to factor values in previous periods and distinguish a sudden increase in a factor from its consistently high values.

Then, each set of factors enters its own layer of neurons containing 512 neurons, and a function applied to the sum of values is relu (this function turns negative sums into 0 and does not affect positive sums).

Next, factors related to environmental improvement and protection are combined into a common matrix, as they are semantically related. Socio-economic factors, in turn, are combined with investment factors and expenditures on federal programs. This is called "simple concatenation."

Calculations at this stage continue to proceed in parallel for groups of factors. Two combined groups of factors sequentially pass through 3 layers of a neural network (256, 128, 64 neurons each). The functions of these layers are also relu.

Then, the data is combined by simple concatenation and jointly passes through 2 more layers of the network (256 and 64 neurons - relu) to better approximate the target values. After passing the data through all the layers and functions of the network, the prediction error was calculated (real values - numbers obtained by the network). The function for calculating it is called the loss function. Model Testing Description. The model contains 1,250,468 parameters for adjustment. This allows for a very detailed and efficient approximation of the target vector function [12].

The loss function calculates the difference between real and predicted values. The most popular loss function for regression tasks (predicting numerical values) is the mean squared error (MSE):

$$L(y, \hat{y}) = \frac{1}{n}\sum_{i=1}^{n}(y - \hat{y})^2 \tag{1}$$

where y – represents the real values;

$\hat{y}$ – represents the predicted values;

n – is the number of elements in the sample.

Then, the weights of the network synapses were updated using the partial derivatives of the overall loss function with respect to each weight. Gradient methods were employed for weight updates. The gradient of the loss function with respect to the weights was calculated using backpropagation. The weights were updated by subtracting from their current values the product of the partial derivative value and the optimizer's value (a special function whose value can be either constant or dependent on the current derivative value). In this task, the Adam optimization function was used.

$$w_t = w_{t-1} - \eta\frac{m_t}{\sqrt{v_1 + \in}} \tag{2}$$

where w_t – is the weight value at the current step;

w_{t-1} – is the weight value at the previous step;

η– is the learning rate;

m_t – is the bias-corrected running average of the gradient;

v_t – is the bias-corrected estimate of the squared gradient;

$\in$ – is a small positive number to avoid division by zero.

Adam is one of the most popular optimizers that combines the advantages of two other methods: AdaGrad and RMSProp. The weight update formulas using Adam are more complex than simple gradient descent. However, the main idea is that Adam adaptively adjusts the learning rate for each parameter based on historical gradients [13].

The essence of the Adam algorithm lies in combining the ideas of two other optimizers: Momentum (takes into account previous gradients to smooth the current gradient) and RMSProp (adjusts gradients based on their recent scale).

For training the model, a commonly used regression task employs the Mean Squared Error (MSE) function. As an additional metric, the Mean Absolute Error (MAE) function is utilized. Let's consider a graphical description of the model architecture (Fig. 2).

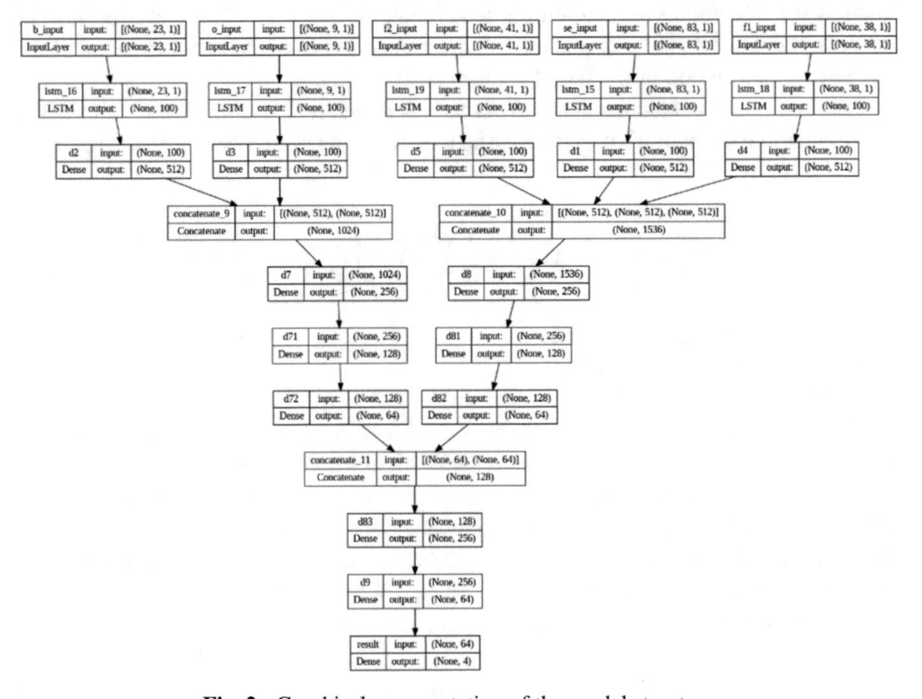

Fig. 2. Graphical representation of the model structure.

Training involved 1000 epochs, at the end of each epoch, target forecast quality metrics were computed, and model parameters were adjusted. Subsequently, the model weights were saved for future use in predictions.

Let's examine how the model was trained on Fig. 3. Small fluctuations in the loss function are visible during parameter tuning.

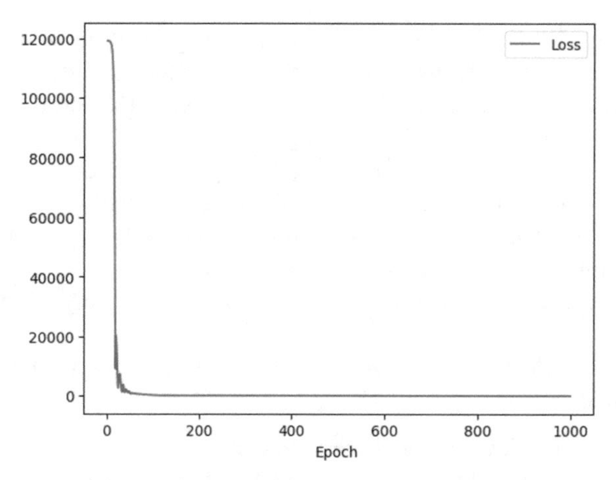

Fig. 3. Plot of the model training process.

Let's also examine the plot of the Mean Absolute Error metric across epochs (Fig. 4).

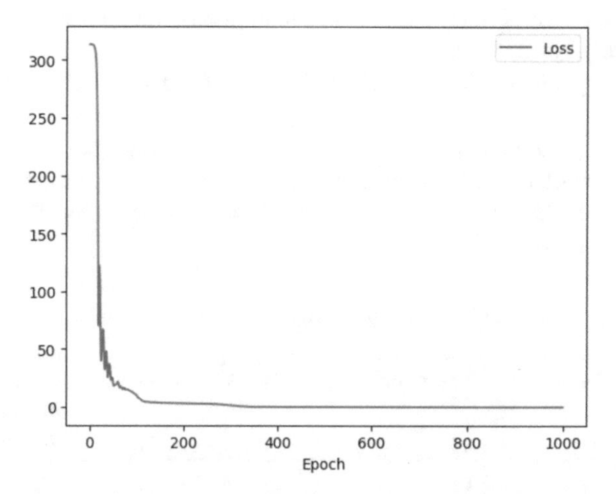

Fig. 4. Plot of the mean absolute error metric across epochs.

In accordance with the Moscow city database, we have developed the FutureAnalytics software. Predicting mortality based on various indicators is a complex and responsible task. The architecture we described is well-thought-out, considering the use of LSTM to account for temporal dependencies and multiple layers to process different groups of indicators. The dataset may vary depending on the specifics of the study, but some features that distinguish LSTM neural networks should be taken into account.

Data quality: The data used must be of high quality, relevant, and up-to-date. The quality of the forecast largely depends on the quality of the input data.

Data dimensionality: Given the large number of layers and neurons, it is important to have enough data for training to avoid overfitting. If there is a small amount of data, regularization or model dimensionality reduction may be needed.

Interpretation of results: Depending on the analysis needs, understanding which factors specifically influence the predicted mortality indicators may be necessary. In this case, deep learning interpretation methods may be useful.

Let's examine the model forecast for known values. It closely aligns with the real values.

Next, we will select negative socio-economic factors. The analysis includes the following factors:

- Number of pregnancies with an abortive outcome;
- Total number of disabled people as of January 1, thousand people;
- Total number of pensioners as of January 1, thousand people;
- Consumer Price Index;
- Changes in the cost of the conditional (minimum) food basket by December of the previous year;
- Changes in the cost of a fixed set of consumer goods and services,
- Housing market price index;
- Emissions of pollutants into the atmospheric air from stationary sources, thousand tons;
- Capture of air pollutants from stationary sources, thousand tons;
- Discharge of polluted wastewater into surface water bodies, million cubic meters;
- Sale of alcoholic beverages to the population, thousand deciliters (Table 1).

Table 1. Model prediction for types of mortality

Options	Infant mortality	Tuberculosis mortality	Cardiovascular	Cancer mortality, including malignant tumors
1	4.8	1.8	490.4	215.3
2	3.5	1.5	516.7	214.1
3	3.6	1.4	510.5	199.3
4	3.6	1.4	510.5	199.3

Now let's create test samples as copies of the data for the last known year of observations. If a factor is positive, subtract 10% from its value, and add 10% to negative factors. The mortality forecast for such a cumulative dynamics demonstrates that simultaneous changes in socio-economic factors according to the conducted scenario lead to an increase in all types of mortality (Fig. 5) (Fig. 6).

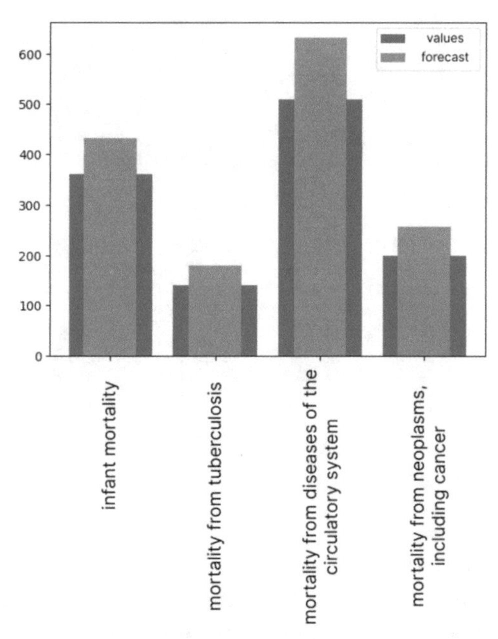

Fig. 5. Forecasted and initial values of mortality types. Next, we will ensure that an increase in only negative socio-economic factors by 20% does not yield the same effect.

Thus, a 20% reduction in negative factors and a 20% increase in positive ones result in a significant decline in all types of mortality. However, increasing the magnitude of some positive factors by 10% alone does not yield the desired effect on mortality levels. This demonstrates that the healthcare system is a complex multilevel system where factors are interrelated, and changing one or several factors individually may have a negative effect despite their apparent usefulness.

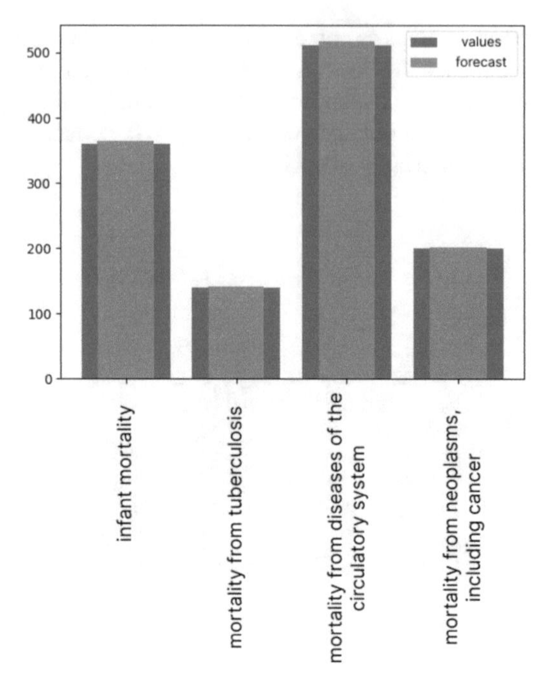

Fig. 6. Change in forecast with an increase in only negative socio-economic factors by 20%.

Healthcare system factors need to be considered collectively; only a balanced influence on multiple factors simultaneously can achieve the desired effect.

As an experiment, let's examine the impact of investments on mortality rates. For this purpose, we will consider the funding for federal projects in 2023. Let's assume that the other factors remain unchanged from 2022. And we note that this did not have a significant impact on mortality, as an increase in funding should be accompanied by the dynamics if socio-economic factors otherwise, it has no effect (Fig. 7).

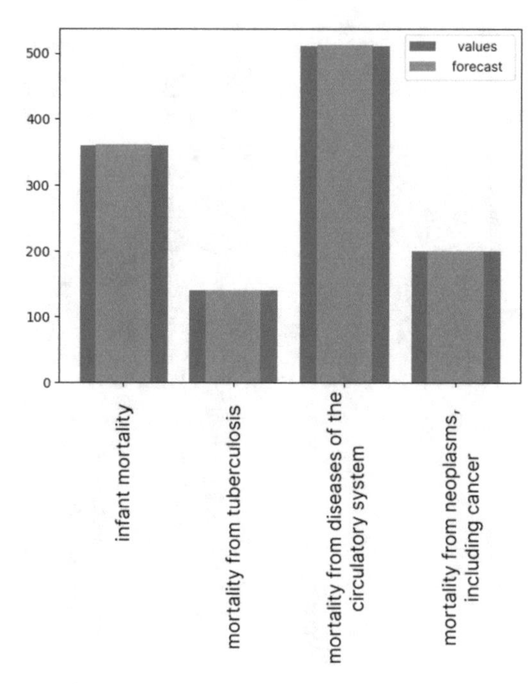

Fig. 7. Mortality change from investments.

Thus, the desired effect of a significant reduction in mortality is achieved by reducing the volume of negative factors by 25% while simultaneously increasing positive factors by 25%, along with a 20% increase in funding and improving urban amenities by 20%.

Figure 8 illustrates the situation where the aforementioned values are achieved. In this case, significant reduction is expected in infant mortality and mortality from circulatory system diseases, with a slight but positive effect on the other two types: mortality from tuberculosis and malignant neoplasms.

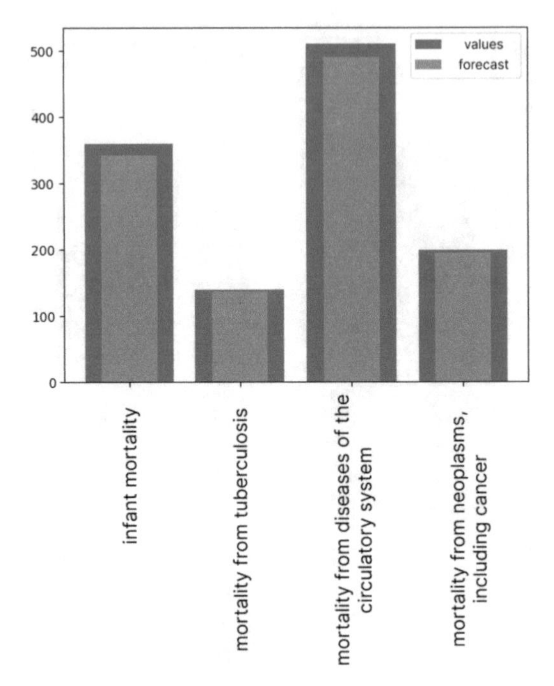

Fig. 8. Reduction in population mortality.

4 Discussion

In the modern world, the problem of healthcare system underfunding is prevalent globally, even in high-income countries. This suggests that the focus should be on improving the quality of investment rather than simply increasing it. Based on this, there is a need for automated systems, such as the one presented by us [14]. Certainly, the forecasted values may undergo changes with the introduction of more specific and narrowly focused data that closely align with mortality indicators.

For a more detailed analysis, more specific data is required. For instance, to study the influence of dietary behavior on health, it's necessary to synchronize information about purchases from mobile banking devices with a program that analyzes it [15]. This would provide a clear picture: a person has been admitted to the hospital with type 2 diabetes and excess weight. Analyzing their food purchases over recent periods reveals excessive consumption of sugary beverages, highly processed foods, and a lack of vegetables. This is a hypothetical scenario that could establish clear connections between factors.

5 Conclusion

In conclusion, we have highlighted the significance of a more qualitative assessment of decision-making in the healthcare sector. The right balance between invested resources and achieved results will lead to a positive trend in reducing the population's morbidity and, consequently, mortality.

With the forecast compiled in the FutureAnalytics program, we were able to identify factors that should be kept under control to ensure a reduction in population mortality.

At the beginning of its journey, the FutureAnalytics program focused on forecasting four types of mortality. Based on the analysis of a multitude of parameters, from socio-economic to medical, the program provided accurate forecasts, enabling informed decision-making in healthcare and social policy.

However, in today's world, the task is not only to understand the current situation but also to see perspectives, adapt, and react promptly to changes. As our understanding of how various aspects of life impact health and well-being expands, the need to integrate new data sources becomes evident.

References

1. Averchenkova, E.E.: Features of assessing the effectiveness of managing a regional socio-economic system from the perspective of management theory. Information and Communication **2**, 7–13 (2020)
2. Andreev, E.M.: Poorly defined and precisely unestablished causes of death in Russia. Demographic Review **2**, 103–142 (2016)
3. Totten, A., et al.: Telehealth: Mapping the Evidence for Patient Outcomes From Systematic Reviews. Rockville (MD). Agency for Healthcare Research and Quality (US) **16**, 667–672 (2019)
4. State Program of the Russian Federation "Healthcare Development", www.rosminzdrav.ru/ministry/programms/health/info, last accessed 13 October 2023
5. ZdravInform: Healthcare System Project, https://zdravinform.info/, last accessed 13 October 2023
6. Showcase of Statistical Data, https://showdata.gks.ru/finder/, last accessed 03 November 2023
7. Healthcare, Federal State Statistics Service, https://rosstat.gov.ru/folder/13721, last accessed 03 November 2023
8. Healthcare in Russia: Statistical Digest. Moscow, Federal State Statistics Service (2021)
9. UHC: Universal Health Coverage, https://www.who.int/ru/news-room/fact-sheets/detail/universal-health-coverage-(uhc), last accessed 03 October 2023
10. Vasin, S.G.: Artificial Intelligence in State Management. Management **3**, 5–10 (2017)
11. Rybakov, D.A.: Development and application of neural networks in various industries. Bulletin of Science **7**(64), 273–277 (2023)
12. Makarenko, A.V.: Deep neural networks: genesis, formation, Current State. Issues of Management **2**, 3–19 (2020)
13. Ivanov, A.N., Mustafina, S.A., Morozkin, N.D.: Algorithm for training neural networks with pseudorandom distribution of connections. Marchukov Scientific Readings **7**, 29–33 (2019)
14. Tarasenko, E.A.: Development of mHealth technological innovations: opportunities for physicians for disease prevention, patients' diagnosis, and counseling. Medical Doctor and IT **4**, 59–65 (2014)
15. Piccininni, C.R.: Cost-effectiveness of robotics and artificial intelligence in healthcare. Unive. West. Ont. Med. J. **87**(2), 49–51 (2018)

Estimation of the Cost of Developing a Digital Twin Using Virtual Reality Technologies in the Metallurgical Industry

Anastasiya Vasilyeva(✉), Margarita Kuznetsova, Ekaterina Zinovyeva, Nina Kuznetsova, and Irina Ageeva

Nosov Magnitogorsk State Technical University, 38 Lenina Avenue, Magnitogorsk 455000, Russia
agvasileva@inbox.ru

Abstract. The relevance of the study is predetermined by the fact that digital twins remain one of the most promising technologies and determine the trajectory of industrial development for the long term. The object of the study is the process of using a digital twin in the metallurgical industry. The work reveals the essence and characterizes the key properties of various types of digital twins – Digital Twin Prototype (DTP), Digital Twin Instance (DTI), Digital Twin Aggregate (DTA), on the basis of which their comparative analysis is carried out. The authors have developed a hierarchical structure for the digital twin work project based on taking into account the processes that are necessary for the successful creation of a digital twin. The work explores and identifies the resources necessary to create a digital twin, as well as their cost. Based on this, the authors calculated the development assessment and substantiated the economic efficiency of a digital twin of a small-section wire mill in the metallurgical industry, which allows the process of testing new technological production modes and their analysis in a day with high accuracy, without stopping the main work and using additional material resources. The results of the study may be useful for specialists of industrial enterprises when forming strategies for the industrial complex in the process of digitalization. The results obtained can become the basis for the further development of the methodology for using the digital twin concept in the metallurgical industry.

Keywords: Digital Twin · Technological Development · Industrial Production · Digital Economy

1 Introduction

Focusing on Industry 4.0 and digitalization processes is associated with assessing the scale of the impact of new technologies on the economic, social and production spheres.

A digital twin is a synchronized virtual model of any objects, systems, people, processes and environments. The digital twin tracks the past and predicts the future. Using a digital twin, you no longer have to stop existing production to test/search for new

A. Gibadullin (Ed.): DITEM 2023, LNNS 942, pp. 66–77, 2024.
https://doi.org/10.1007/978-3-031-55349-3_6

technological solutions, and you also don't have to spend money on materials for tests. Unlike the analytical method of solving, digital twins and their computing capabilities help solve the problem more accurately.

The relevance of the research topic is related to the need to quickly determine optimal solutions for creating new steel grades or new diameters of existing steel grades, as well as improving the production of existing ones.

The goal of the study is to reduce the time spent on researching new metals at "Mill 170".

2 Materials and Methods

The theoretical and methodological basis of the study was the fundamental works of leading foreign and Russian scientists in the field of digital twin implementation [1, 7, 9, 11].

The following research methods were used in the work [13, 14]:

- Firstly, analysis and modeling of the process of testing new technological modes of steel production at "Mill 170";
- Secondly, a set of scientific methods of observation, collection of available information, analysis of experience in the development and application of a digital twin, detailing of its structural elements, as well as assessment of existing and future economic parameters of the digital twin market.

The object of the study is the process of using a digital twin in the metallurgical industry.

The subject of the study is reducing the time spent on researching new metals at "Mill 170".

The practical significance of the study lies in estimating the cost of a digital twin of a small-section wire mill in the metallurgical industry, which makes it possible to carry out the process of testing new technological production modes and their analysis within a day with high accuracy, without stopping the main work and using additional material resources.

3 Discussion and Results

3.1 Feasibility Analysis and Initiation of a Digital Twin Development Project

In 2018, Gartner named digital twins among the leaders for the first time in its annual Technology Cycle Study.

A digital twin is a digital (virtual) model of any objects, systems, processes or people. It accurately reproduces the form and actions of the original and is synchronized with it [2, 10].

A digital twin is a digital copy of a living and non-living physical entity [4]. A digital twin is needed to simulate what will happen to the original under certain conditions. This helps, firstly, to save time and money (for example, if we are talking about complex and expensive equipment), and secondly, to avoid harm to people and the environment.

The concept of a digital twin was first described in 2002 by Michael Greaves, a professor at the University of Michigan. In his book The Origins of Digital Twins, he breaks them down into three main parts:

- Physical product in real space;
- Virtual product in virtual space;
- Data and information that combine virtual and physical products [5, 6].

According to expert estimates, the growth of the global market for digital twins in the near future is planned to average 40% of current figures, and the technology itself ranks second among technologies that guarantee economic development and leadership (see Fig. 1).

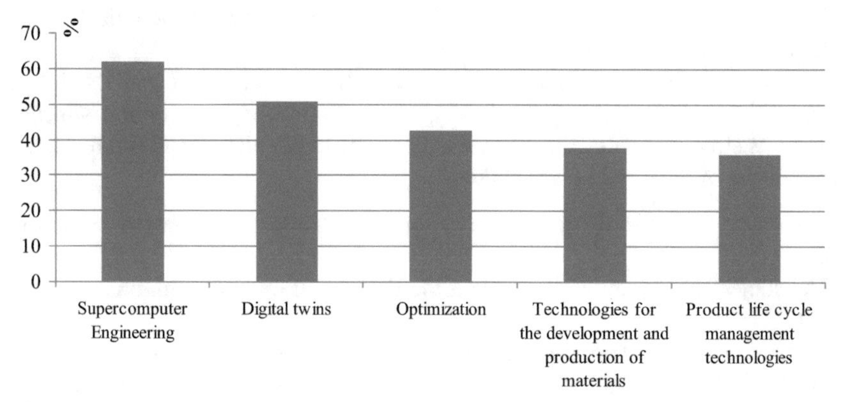

Fig. 1. Top 5 technologies used to increase technological leadership and enter the international market [12].

Digital Twin is a virtual interactive copy of a real physical object or process that helps you manage it effectively, optimizing business operations. For example, a digital twin of a mill can model the location of equipment, the movement of workers, the temperature of steel, the work pattern, and predict the result of the mill's operation.

There are several types of digital twins, namely: prototype (Digital Twin Prototype, DTP), instance (Digital Twin Instance, DTI), aggregated twin (Digital Twin Aggregate, DTA) (see Fig. 2).

DTP is a digital twin that contains product requirements, a three-dimensional model of the object, and a description of process technologies.

Such digital twins are used at Mercedes. They use this type of double to develop and evaluate the design of the machine. This type of double helps them see the car in volume even before its production, evaluate the proportions, and place the lines correctly.

DTI – this digital twin helps to view a 3D model of an object, stores information about materials and components of the product, information about work processes, test results, information obtained from equipment repairs, and data from sensors.

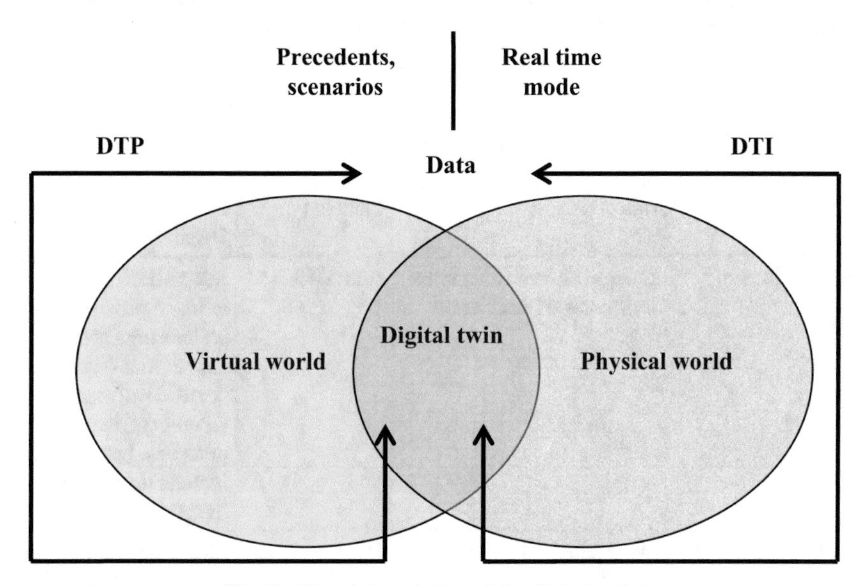

Fig. 2. Visual description of the digital twin.

These digital twins can quite accurately predict a particular production process once the necessary information is obtained. In our case, if the digital twin receives a full set of formulas for how it behaves under certain conditions, then based on data from sensors, test results and information obtained from equipment repairs, it will be possible to obtain a fairly accurate digital twin.

DTA is a computing system of digital twins and real objects that can be controlled from a single center and exchange data internally.

Digital twins have long been used in various fields, such as freight transportation, electrical farms, business, air transport, oil facilities, and healthcare. Greater computing power helps predict certain results with high accuracy.

One of the most famous digital twin companies is General Electric. General Electric is focused on creating digital twins for critical infrastructure in the areas of energy production and distribution, air and locomotive transportation, and healthcare. Digital twins draw on data from maintenance, online sensors, manufacturing and operational data and use a set of high-quality computational physics-based models and advanced analytics to predict the health and performance of an asset over its lifespan, producing highly accurate results. As data is constantly collected to maintain an up-to-date model, the accuracy of the results becomes increasingly higher [8].

GE's digital twins increase in value as they are used, as the algorithms learn more from their impact and become more accurate. They reflect domain knowledge and are an ever-improving repository of intellectual capital. GE expects digital twins to become an integral part of every business. They also say that "The Digital Twin is comprehensive but pragmatic".

To realize the full potential of digital twins, a reliable platform is needed that allows them to live, learn and work on an industrial scale. GE Digital Twins can run on multiple industrial cloud platforms such as AWS and Microsoft Azure [3].

We provide a brief overview of digital twins in Table 1.

Table 1. Comparative analysis of digital twins.

Developer/product	Industry	Double type	Features and Disadvantages
General Electric	Energy, air and locomotive transport and healthcare	DTI, DTA	Integration of digital twins of products and components provides "operational" and "predictive" data Good experience and practices, but does not include the field of metallurgy
IBM Oil	Oil production and oil refining	DTP	Lack of connection of visualization to real process data Information and presentation data
Maserati/Avvocato Agnelli Plant	Automotive industry	DTP	The Siemens CD concept was used. Provides quality data

An analysis of the literature and published cases showed that there are no such solutions for metallurgical production yet or they are at the conceptual stage.

3.2 Planning a Visualization Project for a Digital Twin of a Small Section Wire Mill

According to PMBOK, the process of managing the content of the project described in the materials of the article includes processes that ensure the execution of only those works that are necessary for its successful completion. These are the following processes:

- Content planning;
- Definition of content;
- Creation of ISR;
- Confirmation of content;
- Control of content changes.

These processes interact with each other, as well as with processes from other project management groups. The first three processes belong to the group of planning processes, the other two – to the group of monitoring and control processes.

The input to the “Scope Planning” process includes the results of the initiation group processes – Project Charter, preliminary content of the project description and project management plan.

The Scope Definition process is linked to the Scope Planning process and to the monitoring and control group processes, receiving as input the Project Scope Management Plan and Approved Change Requests.

The Create WBS process is linked to the Define Scope process. The inputs for the “Confirmation of Contents” process are the outputs of the “Creation of WBS” process and the “Management and management of project execution” process of the group of monitoring and control processes.

The Scope Management process is linked to the Scope Validation process and the Monitoring and Document Management Group processes of Performance Reporting and Project Management and Management.

Having determined the approach to the definition, the authors formed a hierarchical structure of work (see Fig. 3).

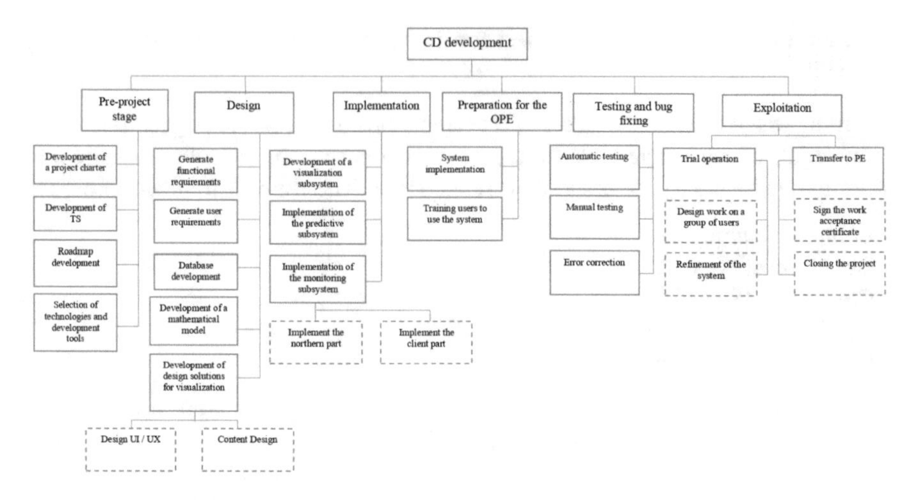

Fig. 3. Hierarchical structure of the digital twin work project.

3.3 Estimation of the Cost of the Digital Twin of a Small-Section Wire Mill

Estimating the cost of a digital twin of a small-section wire mill is based on calculating the labor intensity of the project and determining the financial investments in the project.

Let's calculate the cost of visualizing a digital twin.

At the first step, we will calculate the costs of studying metal processing modes (Table 2).

Table 2. The cost of an hour of work for specialists, taking into account tax rates and insurance contributions.

Specialist position	Wage, ruble/month	Labor costs, ruble/month	Labor costs, ruble/hour
Materials scientist	40,000	52,000	325
Shift Supervisor	100,000	130,000	812.5
Engineer	65,000	84,500	406.25

The research is carried out in two stages. The first stage is rolling. At the first stage, the plant workers (3 engineers and a shift supervisor) roll metal with different rolling parameters. This stage takes on average 4 h. After which the metal is transferred to a laboratory where the researcher studies the properties of the new metal.

Table 3 shows the number of working hours for each of the labor resources at the project implementation stage, and also calculates the total labor costs.

Also, in addition to staffing costs, research requires costs for material and intangible investments (Table 4).

Total costs per rolling (S) $= 2{,}600 + 3 \times 250 + (1{,}625 \times 3) + 190{,}000 = 200{,}725$ ruble.

Table 3. Calculation of labor costs at the project implementation stage.

Project stage/Specialist	Labor costs, hour	Rate, ruble/hour	Labor costs, ruble
Researcher	8	325	2,600
Shift Supervisor	4	812.5	3,250
Engineer	4	406.25	1,625
Total			7,475

Table 4. Calculation of material and non-material costs when researching new properties of metal.

Categories and articles of attachments	Price, ruble
Material investments	
Metal	180,000
Intangible investments	
Electricity	10,000
Total	190,000

On average, 3 studies are carried out per month. The monthly cost of the study is 602,175 rubles/month.

Let's calculate the costs of creating a digital twin of a small-section wire mill.

To create and use a digital twin of a small-section wire mill, we will need two computers and two VR systems (Table 5).

Table 5. Equipment costs for developing a digital twin of a small-section wire mill.

Name	Price for one	Quantity	Sum
SystemVR HTC Vive Pro Eye	183,040	2	366,080
Computer	172,500	2	345,000
Total			711,080

In this case, the amount of depreciation of the computer and VR system, reflecting the process of periodically transferring the initial cost of a fixed asset or intangible asset to production, commercial or general expenses, will be:

Ahtc = (183,040 × 1/3 × 100)/100 = 61,013.34 ruble.
Acomp = (172,500 × 1/3 × 100)/100 = 57,500 ruble.

In addition to computers and VR systems, a development team is required for development (Table 6).

Let's calculate the labor costs for one hour of developers (Table 7).

Table 6. Positions and roles of developers in the project.

Specialist position	Role of the specialist
Analyst	Project management, paperwork
Senior programmer	Digital Twin Programming
Senior designer	Creation of UI/UX interface
Manager	Setting goals
Mathematician	Creating formulas to calculate digital twin metrics
Tester	Digital twin testing

Table 7. Costs for developers of a digital twin of a small-section wire mill.

Specialist position	Wage, ruble/month	Labor costs, ruble/month	Labor costs, ruble/hour
Analyst	42,982	55,876.6	349.23
Senior programmer	36,842	47,894.6	299.34
Senior designer	24,561	31,929.3	199.56
Manager	73,684	95,789.2	598.68
Mathematician	36,842	47,894.6	299.34
Tester	36,842	47,894.6	299.34

Since each developer will not be working on this project all the time, each developer has his own labor costs. Using labor costs, we calculate labor costs (Table 8).

Table 8. Calculation of labor costs at the project implementation stage.

Specialist position	Labor costs, hour	Rate, ruble/hour	Labor costs, ruble
Analyst	58.08	349.23	20,283.28
Senior programmer	130.4	299.34	39,033.94
Senior designer	80	199.56	15,964.8
Manager	63.28	598.68	37,884.47
Mathematician	53.04	299.34	15,876.99
Tester	42	299.34	12,572.28
Total			113,166.5

In order to store backends, it is necessary to provide for the cost of renting server space. Let's calculate the costs of utilities and internet for 3 months. The results of intangible investments in development are presented in Table 9.

Table 9. Intangible investments for the development of a digital twin.

Intangible investments	Price, ruble
Server space rental	10,000
Subscription fee for Internet access	3,000
Public utilities	12,000
Total	25,000

The total cost of developing a digital twin of a small-section wire mill will be S = 849,246.49 rubles.

We will calculate the monthly costs of maintaining the simulator. A system administrator will be hired to accompany the digital twin, so that in case of problems with the PC or VR headset itself, he can fix the problem on the spot (Table 10).

Table 10. Costs of paying employees to support the digital twin.

Specialist position	Labor costs, hour	Rate, ruble/hour	Labor costs, ruble
System Administrator	120	300	36,600

A maintenance agreement will also be drawn up, which will allow you to make the desired changes to the digital twin once a month, as well as complain about bugs and inaccuracies in the operation of the digital twin.

Table 11 reflects intangible investments to support the digital twin.

Total cost of maintaining a digital twin is:

S = 36,600 + 31,300 = 67,900 rub.
Every month the digital twin will pay for itself by
S = 200,725 – 67,900 = 132,825 rubles/month
and depreciation:
S = 132,825 + 61,013/12 + 57,500/12 = 142,701.08
S = 849,246.49/142,701.08 = 6 months.

In total, in 6 months, the money saved by replacing a real study with a digital twin will cover the costs of a digital twin.

Table 11. Intangible investments to support the digital twin.

Intangible investments	Price, ruble
Electricity	700
Internet	600
Support from the developer	30,000
Total	31,300

4 Conclusion

The introduction of digital twins into the production process is currently a key area of digitalization of production. This is due not only to increasing the efficiency of the production process itself, but also to the ability to more optimally use company resources. Digital twins are not universal – there are several types that perform different tasks. DTP is a digital twin containing product requirements, a three-dimensional model of the object, and a description of process technologies. DTI provides the study of a 3D model of an object, stores information obtained from equipment repairs, and data from sensors. DTA is a computing system of digital twins and real objects that can be controlled from a single center and exchange data internally. At the same time, the creation of all digital twins must go through a number of mandatory stages in a certain hierarchy. Digital twins are used in a variety of sectors of the economy, including metallurgy. In the above study, the authors propose to introduce a digital twin into the activities of a small-section wire mill in the metallurgical industry. To do this, the work identifies the directions and amounts of necessary costs – for remuneration of labor for the developers of the digital twin project and its further support, for the purchase of intangible assets for project development, etc. Based on the calculation results, it was revealed that the payback period for the digital twin will be about 6 months. Thus, the economic effect of implementing projects to introduce digital twin technologies is manifested in the effective distribution and use of financial, material, technical, and human resources.

References

1. Abramovici, M., Göbel, J.C., Savarino, P.: Reconfiguration of smart products during their use phase based on virtual product twins. CIRP Ann.-Manufact. Technol. **66**(1), 165–168 (2017)
2. Bolton, R.N.: Customer experience challenges: bringing together digital, physical and social realms. J. Serv. Manag. **29**(5), 776–808 (2018)
3. Borovkov, A.I., Ryabov, Y.A., Maruseva, V.M.: A new paradigm for digital design and modeling of globally competitive products of the new generation. Digital production. Methods, ecosystems, technologies (2018)
4. El Saddik, A.: Digital twins: the convergence of multimedia technologies. IEEE MultiMedia **25**(2), 87–92 (2018)
5. Grieves, M.W.: Digital Twin: Manufacturing Excellence through Virtual Factory Replication – LLC, 7 (2014)
6. Grieves, M.: Origins of the Digital Twin Concept. Montreal 226–242 (2016)

7. Klingenberg, C.O., Borges, M.A.V., Vale Antunes, J.A.: Industry 4.0: what makes it a revolution? A historical framework to understand the phenomenon. Technol. Soc. **70**, 102009 (2022)
8. Official website of the company General Electric. https://www.ge.com/. Last accessed 20 Nov 2021
9. Rad, F.F., et al.: Industry 4.0 and supply chain performance: a systematic literature review of the benefits, challenges, and critical success factors of 11 core technologies. Ind. Market. Manag. **105**, 268–293 (2022)
10. Tao, F., et al.: Digital twin-driven product design framework. Int. J. Product. Res **57**(12), 3935–3953 (2018)
11. Tao, F., Zhang, H., Liu, A., Nee, A.Y.: Digital twin in industry: State-of-the-art. IEEE Trans. Ind. Inf. **15**, 2405–2415 (2019)
12. Yudina, T.N.: Digitalization as a trend in the modern development of the economy of the Russian Federation: pro y contra. State and municipal administration. Scientific notes of SKAGS 3, 139–143 (2017)
13. Zinovieva, E.G., Koptyakova, S.V.: Assessment of integration risks for metallurgical enterprises using the fuzzy set method. CIS Iron Steel Rev. **17**, 58–64 (2019)
14. Zinovyeva, E., Kuznetsova, M., Kostina, N., Vasilyeva, A.: Reducing the international risks of russian industrial companies based on the transfer to the IBM planning analytics platform. Lect. Notes Netw. Syst. **432**, 27–38 (2022)

Formation of Artificial Collective Intelligence Based on the Theory of Active Systems

Nikolay N. Lyabakh[1], Maksim Bakalov[2(✉)], Vyacheslav Zadorozhniy[2], and Yulia Bakalova[2]

[1] Maykop State Technological University, Maikop, Russia
[2] Rostov State Transport University, Rostov-On-Don, Russia
Maxim_bmw@mail.ru

Abstract. The object in this paper is the process (its content, structure, dynamics) of formation of artificial collective intelligence at a variety of enterprises united by socio-economic and/or production relations. Production clusters and self-regulating organizations (SRO) are identified as forms of such relations. The research of concepts related to the category of "intelligence" has been carried out. It is substantiated that the tools for organizing and functioning of clusters and self-regulating organizations should be more actively used in the formation of transport processes. In particular, it is shown that transport and logistics chains of freight delivery have both the properties of clusters and the properties of self-regulating organizations. The mechanism of formation of transport and logistics chains of freight delivery is considered. The spectrum of mathematical tools for the synthesis of artificial intelligence systems is analyzed. In particular, the theory of pattern recognition, fuzzy sets, games, percolation, active systems. It is substantiated that the models and methods of the theory of active systems, designed to stimulate the joint work of enterprises and coordinate their divergent economic and production interests, are an effective tool for the formation of collective intelligence of a variety of interacting agents (enterprises and top-level management structures). The advantages and disadvantages of the basic mechanism for coordinating conflicting interests of economic agents are analyzed. The author's results of the development of basic models are described regarding: a) compensating for the inadequacy of models in practical applications, b) expanding the scope of methods application. In particular: according to the point a) the requirements for the quadratic of the cost function, the absence of initial production costs, and the full sale of manufactured products have been removed; according to point) models for sequential and parallel operation of enterprises in clusters and self-regulating organizations have been clarified. New directions of the research on the topic are substantiated: taking into account several criteria for the quality of enterprises operation and the multidimensionality of influencing factors, checking the adequacy of criterion models (the adequacy interval should be wider than the area of agreed solutions), organizational basis for the formation and use of the obtained collective intelligence.

Keywords: Artificial Collective Intelligence · Clusters · Self-Regulating Organizations · Transport and Logistics Chains · Theory of Active Systems

A. Gibadullin (Ed.): DITEM 2023, LNNS 942, pp. 78–89, 2024.
https://doi.org/10.1007/978-3-031-55349-3_7

1 Introduction

The "intelligence" category, despite its active development in recent years, is studied insufficiently. There are many reasons for this phenomenon:

- Intelligence is a complex concept that includes various properties of individuals and their communities (biological, psychological, knowledge-based). It is obvious that IQ tests developed to assess intelligence do not reflect its essence. As soon as we change the share of the mental, knowledge-based component in their structure and the scope of analysis, as the result of the assessment will change dramatically.
- Intelligence is studied and applied in many scientific fields and practical disciplines (psychology, medicine, technology, mathematics, etc.), which impose their own peculiarities on the understanding of this term.
- "Intelligence", "artificial intelligence" (AI), "intellectual activity" are new actively developing categories, therefore, at each stage of such research, additional meaning is introduced into them.

Thus, we will not further discuss the definition of the category "intelligence", but accept it in the generally accepted conceptual sense, which does not affect the essence of the task set in the research: the formation of artificial collective intelligence based on the theory of active systems.

Intelligence can be classified into individual intelligence and collective intelligence. The relationships between them are quite complex. The intelligence of the collective is not the sum of the intellects of its members (the principle of superposition is obviously not fulfilled here). When holders of private (individual) intelligence interact with each other, there are always synergetic processes that strengthen or weaken the intelligence of the group. So, in psychology, the phenomenon of "crowd intelligence" is well known – it is always lower than the intelligence of the weakest individual in this regard.

There are also opposite examples in nature. We generally deny the presence of intelligence to many "our smaller brothers" (a single ant, for example), but by uniting in a colony, the ants show a pronounced collective intelligence.

Indeed, many objects and processes created by ants clearly indicate that they have a collective mind:

- The structure of an anthill with a "maternity place" and a "kindergarten", "vegetable gardens", food storage, waste places, well-functioning ventilation and insulation in the dwelling.
- Coordinated processes of offspring reproduction in the colony and preservation of the species in nature.
- Organization of dwelling and territory protection, etc.

Like humans, ants, for example, have created technologies for fermentation of products (of plant origin) and long-term storage (nectar, protein food). They pass these recipes from generation to generation, and implement them with precision at the right time and under changing conditions of existence. Isn't this an intellectual activity?

In the theory of artificial intelligence, such concepts as "swarm intelligence" [1], "grey wolf algorithm" [2], "bat algorithm" [3], etc. have already taken root.

This paper is devoted to the development of a similar technology (formation of group intelligence) on a set of agents (people, enterprises) in various fields of activity using the theory of active systems (TAS) [4].

Numerous models (and corresponding examples) of stimulating participants of collective work are given in [5], which clearly demonstrate the non-traditional role of mathematics - the cultivation of a new corporate culture of agents' communication among themselves. To manipulate participants (bribery and/or pressure) and data (distorting them in order to obtain a more advantageous position: a better order, more profit, etc.) becomes unprofitable.

The first purpose of the research is to systematize the types of socio-economic and production facilities to which TAS is further applied. Our basic concepts are clusters and self-regulating organizations (SROs). Using these concepts, we will define below the objects of our research: transport and logistics chains (TLC) for the delivery of goods and passengers [6].

Clusters are groups of enterprises that produce various products within the framework of one, more significant project. For example, they are enterprises that produce body parts, chassis, engines, tires, glass, headlights and other parts for a car [7].

SROs are groups of enterprises that produce the same products, provide the same services, but cooperate in their efforts to avoid ineffective competition, improve the quality, predictability and stability of their work [8]. A typical example is construction enterprises merged into SROs.

It is obvious that the interests of enterprises in clusters and SROs are implemented in various ways. Indeed, in clusters we stimulate a decrease or increase in the output of the corresponding enterprise by "adjusting" it to common goals. And in the SRO, enterprises undergo a filter on the quality of their products, technical and technological equipment (inappropriate enterprises are not allowed into the SRO). The SRO bears collective responsibility for the output of products. Enterprises in SROs can count on participating in projects for which they do not have their own resources.

In these examples, the "intelligences" of the heads of enterprises included into clusters and SROs, their decision-making systems, and the collective intelligence that arises during communication are closely intertwined.

TLCs demonstrate the properties of both clusters (different transport agents complement each other when implementing door-to-door technology) and SROs (different means of transportation can be used along the same route). Clusters and SROs in transport allow avoiding costly duplication, ensure optimization of interaction, improve the quality of transport services, and reduce costs. Therefore, the tools for organizing and functioning of clusters and SROs should be more actively used when forming transport processes.

TLCs contain a group of agents connected both in series (Fig. 1) and in parallel (Fig. 2).

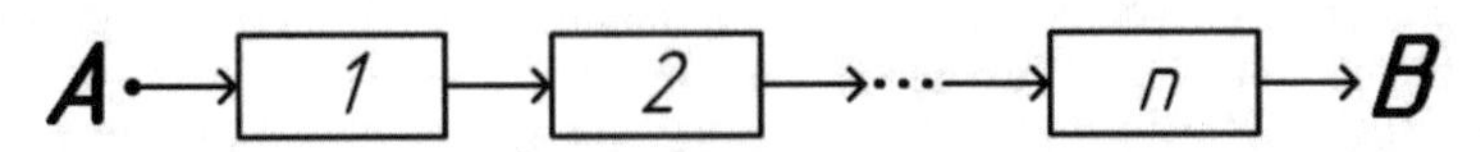

Fig. 1. Series-connected agents carrying out transportation of goods.

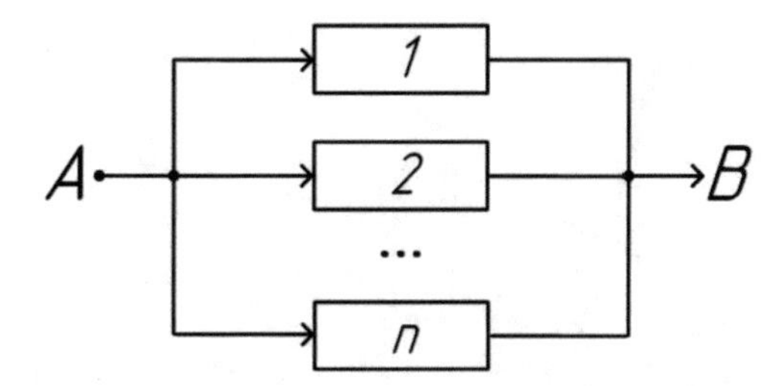

Fig. 2. Parallel connected agents carrying out transportation of goods.

In general, these types of connections are combined. Figure 3, for example, shows such a TLC freight transportation along the route "Mineral'nye Vody – Novorossiysk".

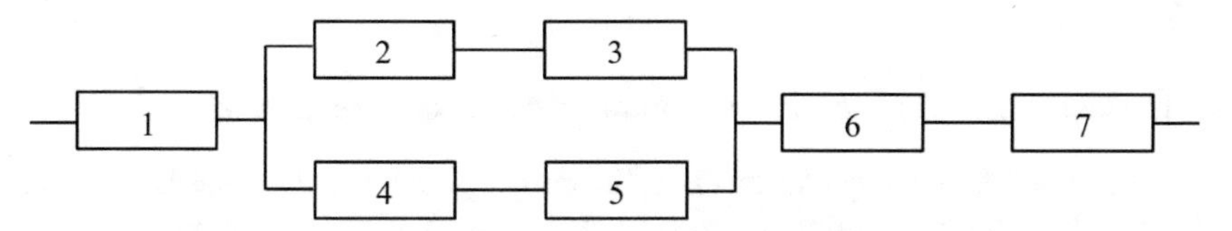

Fig. 3. TLC freight transportation along the route "Mineral'nye Vody – Novorossiysk".

This scheme takes into account two possible ways of delivering freight from Mineral'nye Vody to Novorossiysk (see Fig. 4) having both common and different traffic sections:

- Mineral'nye Vody – Nevinnomysskaya – Kavkazskaya – Krymskaya – Novorossiysk;
- Mineral'nye Vody – Nevinnomysskaya – Belorechenskaya – Goryachiy Klyuch – Krymskaya – Novorossiysk.

The second purpose of this research is to systematize the mathematical tools for AI synthesis and to reveal the possibilities of the mathematical tools of the TAS in the formation of collective intelligence while coordinating the economic interests of agents with unconditional performance of the production functions of the TLC: freight transportation along the route in a given volume and within a specified time.

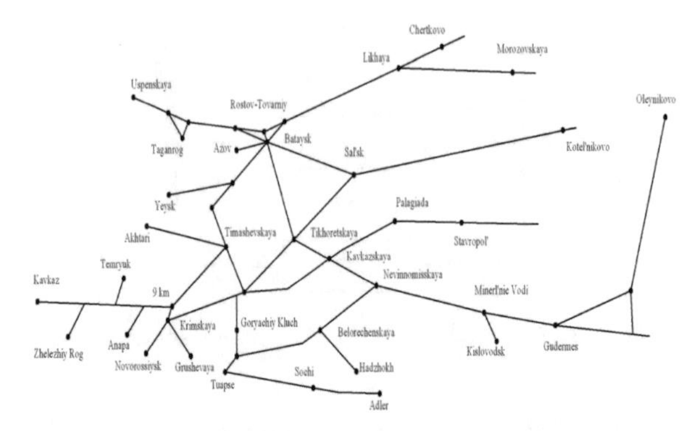

Fig. 4. Fragment of the Southern polygon of Russian Railways.

2 Methods

2.1 Mathematical Tools for the Synthesis of Artificial Intelligence

- The theory of pattern recognition (TPR) is probably the most recognizable tool in this field. Indeed, the recognition of faces, symbols (texts), situations has already become our everyday and easily solved tasks.
- Fuzzy sets theory (FST), operating with linguistic variables and fuzzy concepts, copying the logic of human reasoning, allows to best modeling of human natural intelligence and building up new, similar constructions of artificial intelligence.
- In game theory we would like to distinguish two aspects: a) modeling of human behavior in solving cooperation problems (games with Nature, antagonistic games, coalition games) and b) modeling of human reflexion – one of the key properties of natural intelligence.
- The percolation theory having many technical applications in which the transition processes from quantity to quality are described mathematically, can be extremely useful also in describing the intellectual process of illumination- insight.
- Paying tribute to the above-mentioned theories, let us note TAS as a theory of the formation of collective intelligence of a group of agents interacting on a certain platform. These agents can be people, their individual teams (enterprises), devices and equipment (in particular, robots), decision-making systems and other artificial objects. Further research is devoted to this theory.

2.2 The Basic Mechanism for Coordinating the Interests of Agents [4, 5]

The fundamentals of TAS were developed by domestic scientists of the V.A. Trapeznikov Institute of Control Sciences (ICS). The problem statement is as follows: there is a group of enterprises interconnected by production and/or economic ties, each of which has its own interest that in general does not coincide with the interests of other participants and the group as a whole. It is necessary to develop a mechanism to ensure fair interaction and obligatory implementation of the overall plan.

The essence of the method:

For a cluster enterprise or SRO, hereinafter referred to as an economic entity (EE), a cost function is assumed depending on the volume x of output (services provided) in the form of:

$$z = ax^2 \tag{1}$$

Assuming the full realization of the product at price c, we obtain a profit of EE:

$$P = c \cdot x - ax^2 \tag{2}$$

The maximum profit P (interest of EE) is achieved at the point:

$$x_0 = \frac{c}{2a} = c \cdot e \tag{3}$$

In (3) it is indicated: $e = \frac{1}{2a}$, which is called the EE efficiency coefficient.

If EE manages to receive an order in the amount of x_0, then its revenue will be maximum:

$$y_0 = y(x_0) = 0.5 \cdot c^2 \cdot e \tag{4}$$

Dependence (2), coordinates x_0 and y_0 are represented by an external parabola in Fig. 5.

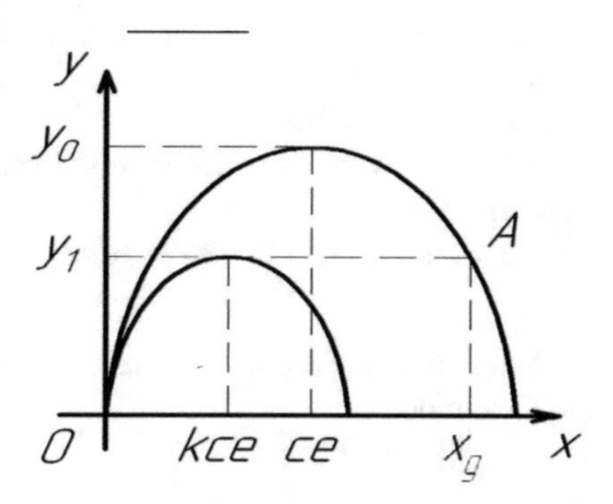

Fig. 5. Geometric interpretation of the synthesis of the basic criterion of the enterprise's activity.

The interests of the higher management body (cluster, SRO) and individual EE do not generally coincide. For example, the EE needs to perform a load in an amount greater than *ce*. In this case, a contradiction arises between the interests of the designated production participants: the enterprise does not "want" to fulfill the top-down plan. It will need additional capacities and resources, which requires additional costs and in this case, it receives less profit.

In [4, 5], to control the EE by the upper management level, it is proposed to reduce the price for the services of the EE if it fails to fulfill the plan. In this case, instead of c, the price will be equal to $c_1 = k \cdot c$, где $0 < k < 1$.

Of particular interest is the point x_g on the axis OX, in which the equality holds (point A in Fig. 1):

$$cx_g - \frac{1}{2e}x_g^2 = \frac{1}{2}k^2c^2e. \tag{5}$$

The point x_g is the abscissa of the point A – the intersection of a large parabola with a horizontal line with an ordinate equal to y_I. This is the boundary point. If the x_z plan of EE satisfies the condition $c_e \le x_z \le x_g$, then it is advantageous to fulfill it, since it will be greater than the maximum possible if the plan is not fulfilled although the revenue will be less. With a plan $x > x_g$, it is beneficial not to perform the plan, but to carry out a load in the amount of x_I, in order to guarantee a profit in the amount of y_I. From (5) it is easy to find x_g (this is the largest root of this quadratic equation):

$$x_g = c \cdot e \cdot \left(1 + \sqrt{1 - k^2}\right) \tag{6}$$

Segment:

$$\left[ce; x_g\right] \tag{7}$$

is called the agreed area.

Limitations of this model:

- A priori, a quadratic cost function is assumed for the production of output (in our case, the provision of transport service).
- All manufactured products are sold in full.
- The production of goods is not associated with the initial costs (the criterion curve passes through the origin in Fig. 4).
- The price of products (services) is constant (fixed) and does not depend on the volume sold.

These restrictions are rather strict and, as a rule, are not fulfilled in practice. This required the development of the model.

3 Results

3.1 The Realized Development of the Mechanism for Coordinating Conflicting Interests of Agents [8, 9]

The development of the mechanism for coordinating conflicting interests of agents was carried out in three aspects: the removal of restrictions in the general formulation of the problem, the refinement of the mechanism under research for series and parallel interaction of agents.

The removal of some restrictions is implemented as follows:

1. The type of cost function for EE is not assigned a priori (as it was done in (1)). The criterion of the enterprise's activity is not formed from the subjective assumptions of the decision-maker (DM), but is calculated based on statistical data in the vicinity of the optimum, objectively characterizing the activity of the corresponding EE. A quadratic dependence was chosen as the approximating polynomial. We obtain expressions of the form presented in Fig. 6 (and not Fig. 5).

$$y = a0 + a1x + a2x2. \tag{8}$$

   Let us transform expression (8) by pointing out a full square in it:

$$y = -mi(x - -ai)2 + bi, \tag{9}$$

   where $i = 1, 2, \ldots, n$.

   Here, the parameters of the model $a > 0$ and $b > 0$ have a well–known economic meaning: a is the optimal value of the enterprise load, at which the maximum revenue equal to b is achieved.
2. The work of EE and their groups is analyzed:
   - if the EE fulfills the plan set by the cluster or SRO, then its remuneration is carried out according to the dependence (9);
   - if the EE does not fulfill the plan, sanctions are imposed on it by proportionally reducing the value of b (penalties) with a coefficient of $0 < k < 1$. That is, if the plan is not fulfilled, the enterprise operates according to the model:

$$y = -mi(x - ai)2 + kibi. \tag{10}$$

3. Let us calculate ratios for key parameters of the procedure: k and x_g [9]:

$$xgi = ai + (bi(1 - ki)/mi)0, 5, \tag{11}$$

$$ki = 1 - mi(xgi - ai)2/bi. \tag{12}$$

   Obviously, as long as the value of the plan belongs to the interval $[a_i; x_{gi}]$, it is profitable for the enterprise to fulfill this plan. If the top-level management assigns a plan $x > x_{gi}$, then it is profitable for the enterprise to carry out freight transportation in the volume of $x_i = ai$ and receive revenue in the amount of $k_i b_i$.

That is, the interval $[a_i; x_{gi}]$ is the area of agreed decisions of the top-level management and the i-th EE.

For the series connection of agents in TLC, the development of the above method was carried out in the paper [8].

- We set n – the number of EE involved in the transport and logistics process.
- Enter n-lines of parameters: m_i, a_i, b_i.
- Set the plan p, which should be implemented by all of this TLC.
- In the cycle we solve n-equations of the form: $-\frac{m(p-a)^2}{b}$. And we get the required n-values of k_i.

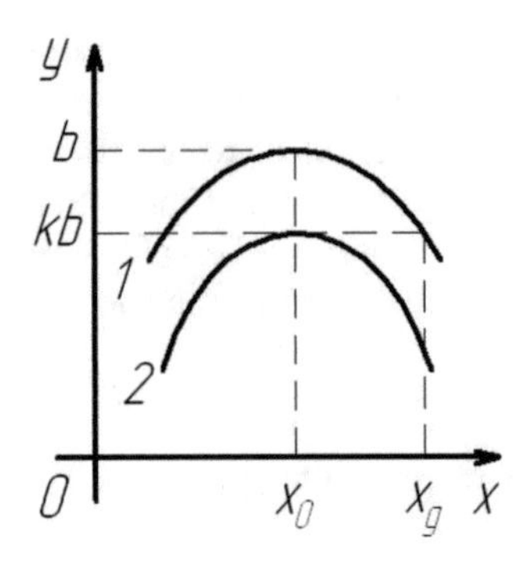

Fig. 6. Generalization of the mechanism for coordinating interests of agents at different levels of management

- We calculate the profit formulas for each EE, taking into account the penalty coefficients.

Generalization of the mechanism for coordinating the interests of parallel working agents of the same management level. For the parallel connection of agents in TLC, the development of interests matching method was carried out in the paper [9]. It is necessary to determine a fair distribution of the load between them with the condition of the general work plan implementation.

Now we have *n* parallel operating channels described by equations of the form (10), (11) and (12). The fairness of the task distribution between the chains requires that $k_i = k$. For practice, the case is interesting when the load *C* on a group of EE exceeds their total interests:

$$C > \sum 1n\ ai. \tag{13}$$

In this case, it would be logically correct (rational) to require equality of the coefficient *k* for all EE operating in parallel. The boundaries of the permissible planning intervals x_{gi} will be different.

So, to determine the lower bound of the coefficient *k*, we require the fulfillment of the equality:

$$C = \sum 1n\ xgi \tag{14}$$

We obtain a system of $n + 1$ equations: *n* equations of the form (11) and one Eq. (14). Solving this system under the condition $k_i = k$, we obtain the required values of *k* and x_{gi}.

4 Discussion

Both in the works of ICS RAS scientists [4, 5] and in the studies of RSTU scholars [9, 10], a one–dimensional case was considered: one quality criterion characterizing the activity of the agent under research and one factor of production taken into account that is the intensity of the enterprise's operation. In this case, there is no contradiction within the enterprise, and the area of agreed (permissible) solutions represented a straight-line segment.

In practice, it is often necessary to focus on several criteria and several production factors. Let us consider two such cases.

1. The enterprise is characterized by several indicators, for example, profit, revenue, profitability.
2. The enterprise is characterized by one indicator, but takes into account several factors of production x_i that is, the criterion is a function of several arguments. In this case, we obtain a multidimensional region of permissible solutions, see Fig. 7 (in Fig. i = 2). To select the optimal point in this area, an external criterion is set, which can have the following form:

   - minimizing the average risk of going beyond the permissible range of parameters;
   - minimizing the maximum risk (otherwise: finding a guaranteed result).

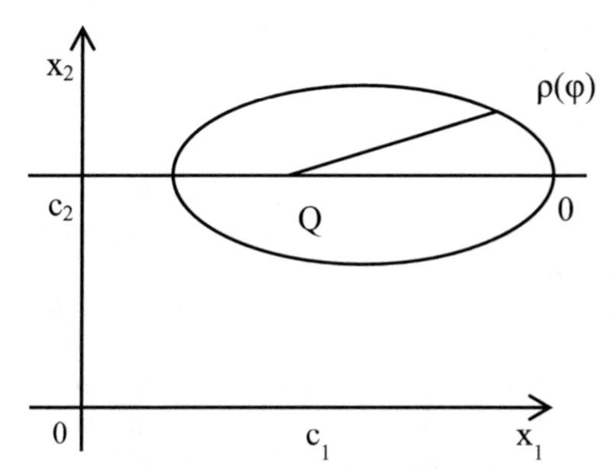

Fig. 7. Setting the permissible range of system parameters (as well as control parameters) in polar coordinates.

For criterion A), its analytical task will have the form:

$$Q_{opt1} = \arg\max \int_0^{360} \rho(\phi)d\phi. \tag{15}$$

For criterion B) respectively:

$$Q_{opt2} = \arg\max\min \rho(\phi). \tag{16}$$

The point found in this way has a certain degree of reliability: the risk of leaving the zone of comfortable functioning is minimized.

Choosing one approach or another is tantamount to choosing the psychological type of the virtual agent (equivalent to how people may be prone to more or less risk).

3. Checking the adequacy of the model characterizing the quality criterion. It is obvious that the used quadratic optimization will have a limited zone of adequacy of the model and it is important that the region of agreed solutions is located inside it.

Organization of agents' interaction.

To organize the interaction of agents generating collective intelligence, a digital platform (DP) is being created with access to the necessary databases (DB) and knowledge bases (KB). The DP provides:

– regulated access of participants to information;
– data transfer using blockchain technologies, which guarantees the safety and accuracy of the information used;
– implementation of the Internet of Things, which are agents, databases and knowledge bases.

The DB accumulates data on the functioning of agents to build dependencies of criteria on production factors.

The KB stores the attracted and acquired knowledge: programs for calculating dependencies and the dependencies themselves.

5 Conclusions

1. The necessity of creating clusters and SROs in transport that improve the quality of transport services, expand their range, and improve the management of transport processes is substantiated.
2. The concepts of "cluster interactions of agents", regulating the volumes of various output between them, and "interactions of SRO agents", which integrate and optimize the resources required to produce one type of product have been introduced. In terms of enterprises, the former are characteristic of clusters, the latter of SROs. These concepts attract various mechanisms for the formation of collective intelligence based on the application of the theory of active systems.
3. The basic and developed mechanisms mentioned above are analyzed.
4. Three problems of prospective research (expanding the areas of practical application of methods and their clarification) are formulated.
5. The digital platform for organizing the interaction of agents generating collective intelligence is commented on.

References

1. Miller, P.: Swarm intelligence: ants, bees and birds can teach us a lot. National Geographic. Russia **8**, 88–107 (2007)
2. Population optimization algorithms: Optimization by a Pack of Grey Wolves (Grey Wolf Optimizer - GWO). https://www.mql5.com/ru/articles/11785. Last accessed 3 Oct 2023
3. Grinchenkov, D.V., Mokhov, V.A., Pivovarov, S.A., Romanov, L.L.: A variant of the implementation of the swarm algorithm of bats. Bulletin of higher educational institutions. North Caucasus Region. Tech. Sci. **4**, 22–27 (2015)

4. Burkov, V.N., Novikov, D.A.: Theory of active systems (history of development and current state). Control Sci. **3**(1), 29–35 (2009)
5. Novikov, D.A.: Theory of management of organizational systems. MPSU, Moscow (2005)
6. Savin, G.: The smart city transport and logistics system: Theory, methodology and practice. Upravlenets **12**(6), 67–86 (2022)
7. Vladimirov, Y.L., Tretyak, V.P.: On classifications of enterprise clusters, Downloads/o-klassifikatsiyah-klasterov-predpriyatiy.pdf. Last accessed 3 Oct 2023
8. Shevchenko, O.M., Efimtseva, T.V., Sweet, Y.: Self-regulation of entrepreneurial and professional activity: unity and differentiation. Infra-M, Moscow (2015)
9. Chislov, O., Lyabakh, N., Kolesnikov, M., Bakalov, M., Zadorozhniy, V., Khan, V.: Intellectualization of logistic interaction of economic entities of transport and logistics chains. In: Beskopylny, A., Shamtsyan, M. (eds.) XIV International Scientific Conference "INTERAGROMASH 2021." LNNS, vol. 246, pp. 369–377. Springer, Cham (2022). https://doi.org/10.1007/978-3-030-81619-3_42
10. Lyabakh, N.N., Bakalov, M.V., Shapovalova, Y.V.: Ensuring the economic security of economic entities of various management levels through the development of a procedure for reconciling conflicting interests. In: XXVII International Conferences, pp. 142–150 (2019)

Nonlinear Modeling of the Population State of Gregarious Locusts

D. K. Muhamediyeva[1(✉)], A. Kh. Madrakhimov[2], and N. S. Mirzayeva[1]

[1] Tashkent University of Information Technologies named after Muhammad al-Khwarizmi, 108 Amir Temur Street, Tashkent, Uzbekistan
dilnoz134@rambler.ru

[2] Academic lyceum under the Tashkent University of Information Technologies named after Muhammad al-Khwarizmi, 108 Amir Temur Street, Tashkent, Uzbekistan

Abstract. The purpose of this study is to analyze the ecology and fauna of locusts in a certain region, as well as to model the status of the population of gregarious locusts. The study includes an analysis of the characteristics of the ecology and fauna of locusts in the region, as well as a study of the properties of mathematical models of a non-linear population of gregarious locusts. The main methodology of the work is computer modeling of the processes of multicomponent cross-diffusion systems of the biological population of gregarious locusts, using the principles of comparison of solutions. To do this, it is expected to find initial suitable approximations, including using the linearization method. Additionally, the research involves the development of algorithms and programs for conducting a computational experiment. As part of the experiment, the correctness and effectiveness of the results obtained will be checked. Visualization of the solution will allow you to more clearly present the results and evaluate the quality of the solutions obtained. Therefore, this study aims to gain a better understanding of locust population dynamics and to develop numerical schemes and methods to model and control these processes.

Keywords: Biological population · Differential Equation · Locust Herds · Nonlinear Models · Algorithm · Program

1 Introduction

In the world, one of the problems in agriculture is the reduction in crop yield caused by the action of pests, diseases, and weeds. Agriculture loses up to 40% of the yield annually. The use of integrated pest management, which involves a combination of rational practices and control measures based on information about insect pests, can solve this problem. The effectiveness of such protection depends on the quality of monitoring insect pests, including modeling the population dynamics of locusts. Targeted scientific research is being conducted in this direction, such as the creation of nonlinear mathematical models for reaction-diffusion in natural sciences, the development of economical numerical schemes for the nonlinear process of the biological population of locusts [1–5].

A. Gibadullin (Ed.): DITEM 2023, LNNS 942, pp. 90–105, 2024.
https://doi.org/10.1007/978-3-031-55349-3_8

In the Republic, certain results have been achieved using a nonlinear splitting algorithm depending on the dimensionality of space. Modeling and studying the processes of competing biological populations are important tasks in ecology and biology. It can aid in understanding population dynamics, predicting and managing population processes, and developing strategies to combat harmful insects, such as locusts. Various methods can be applied for modeling competing biological population processes, including modeling based on differential equations, mathematical biology methods, and others. Creating accurate and efficient mathematical models, as well as developing algorithms for their solution, are important tasks to achieve an understanding of the dynamics of biological systems and the development of methods for their control and management [6–8].

The Kolmogorov-Fisher equation describes the probability that a population will transition from one state to another in the next time period, given that it is in a certain state at the current moment. The model of a single population with nonlinear diffusion, discussed in the work of N.V. Belotelov and A.I. Lobanova, describes the spread of a population taking into account the nonlinear dependence of the migration flow on the local population density. Such dependence may arise, for example, in the presence of competition among individuals for resources. Both models are examples of nonlinear mathematical models of population dynamics and can aid in understanding population dynamics, forecasting, and managing population processes [3–5].

The goal of the research is to analyze the features of the ecology and fauna of the locust region and model the state of the population of gregarious locusts. This involves developing numerical schemes and implementation methods.

The research includes the analysis of the ecology and fauna features of the locust region and the study of the properties of mathematical models of the nonlinear population dynamics of gregarious locusts. It also involves computer modeling of processes in multi-component cross-diffusion systems of the biological population of gregarious locusts based on the principles of comparing solutions.

The objectives further include finding initial suitable approximations, with one method being the linearization method, developing algorithms and programs, conducting a computational experiment to verify the correctness and efficiency of the obtained results, and visualizing the solution to provide a more intuitive representation of the results and assess the quality of the solution.

Locusts pose a significant threat to agriculture in Central Asia and the Caucasus. Three main species of locust pests, such as the Italian locust (CIT), the Moroccan locust (DMA), and the Asian migratory locust (LMI), pose a serious danger to the food security of countries in the region (Figure 1) [9].

Mapping territories is an important stage in locust surveying. It allows identifying locations where insects are found and assessing the degree of infestation. Maps help estimate the approximate area of affected areas and plan pest control measures. Data obtained through mapping enable effective locust control, reducing the negative impact on the environment and the population [10, 11].

In 2020, Central Asian countries faced not only the COVID-19 pandemic and quarantine restrictions but also a massive locust invasion, the largest in the past two decades. Due to the possibility of locusts migrating from neighboring countries, joint monitoring

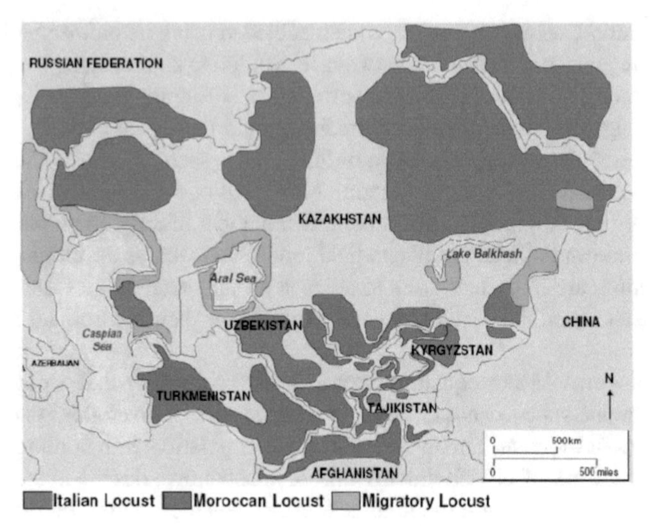

Fig. 1. Geographic distribution of the three main species of locust pests in Central Asia and adjacent regions.

was organized in border areas with the quarantine services of Kazakhstan (May 26–27, 2022) and Tajikistan (March 28-April 2, 2022). During the briefing, it was stated that the situation in border areas was under control.

Experts presented data on the main foci of the spread of the Moroccan locust in the Karshi steppe and the Surkhan-Sherabad Valley. Additionally, issues related to monitoring and population dynamics of other locust species, such as the Italian locust and the large saxaul locust, were discussed [12].

In the Caucasus region and Central Asia, the most critical threat to agriculture is locusts, which can lead to significant economic and social consequences. During outbreaks, locusts can cause serious damage to various crops, including grains, legumes, vegetables, fruits, and vineyards, as well as pastures. Over 25 million hectares of agricultural land are at risk, putting at least 20 million people in danger. Rural communities, where the most vulnerable populations rpeside, suffer the most from locusts and the negative consequences of combating them, leading to potential health and environmental deterioration [13, 14].

Obtaining accurate data on the harm caused by locusts is challenging because many countries provide information only about cultivated and treated lands, omitting disruptions in agriculture and population impact. Very limited information is available. For example, during the 1999 outbreak of the CIT locust in Kazakhstan, 220,000 hectares of grain crops were destroyed, resulting in approximately $15 million USD in damages [15].

Small farmers engaged in natural farming are particularly vulnerable to the threat posed by locusts, as even a small outbreak can cause serious damage and threaten their survival. This issue is especially relevant in the Caucasus, where there is a prevalence of small landowners rather than a few large farms, and it is even more pronounced than in Central Asian countries. For example, during the last locust outbreak in Armenia in 2020, damages affected 70 to 96% of cultivated lands, including grain crops, fruit orchards, and pastures. Even with a relatively small planting area occupied by small farmers in the Caucasus compared to Central Asia, the risk associated with locusts remains significant, highlighting the need for appropriate precautionary and preventive measures [11, 12].

Pilot projects using satellite imagery for monitoring locust breeding grounds have been conducted in Uzbekistan, Kazakhstan, and Russia. However, satellite technology is not yet fully integrated into survey methodologies. Recently, national monitoring programs for pests and diseases of agricultural crops, including locusts, have been launched in Kazakhstan and Russia. Additionally, the Interparliamentary Assembly of the CIS countries is currently working on creating a common satellite monitoring system for pests, which can be beneficial for joint locust control efforts [14]. The main goal is to reduce the frequency and intensity of locust invasions in the Caucasus and Central Asia, as well as to minimize the damage inflicted by locusts on crops and pastures. This will help mitigate the negative impact of locusts on food security and the livelihoods of the most vulnerable rural populations [13].

Certain species of locusts, such as the Asian, Italian, and Moroccan locusts, pose a serious threat. Mass chemical treatments conducted over large areas can lead to the destabilization of the ecological situation, as assessed in this study, which represents a complex system of channels, large and small water bodies, islands, and estuaries (as shown in Fig. 2). The vegetation includes reed beds, as well as other moisture-loving grasses, sedges, and rushes. The soils in these areas are light, sandy, and loamy, sometimes clayey, and occasionally slightly saline.

Fig. 2. In the delta of the Amu Darya River in the Aral Sea region, Karakalpakstan, reed beds serve as breeding grounds for the Asian locust.

For effective plant protection against harmful organisms, accurate and comprehensive information about the distribution, development, economic significance of pests, crop conditions, and the overall environment is essential. Only with timely acquisition and analysis of this information can optimal decisions be made to implement preventive measures and increase their profitability. To monitor and assess the status of pest populations and make decisions on protective measures, systematic information on their harmfulness is required. This creates the need for an information system comprising four main components: information collection, transmission, processing, and storage [16].

In the development of automated forecasting systems, it is necessary to design information support that includes all the required information and determines methods for processing, storing, and presenting it. The development of information support is a complex process that involves solving several tasks, including [16–18]:

Setting System Goals: Before development begins, it is crucial to define what goals the system will achieve, what forecasts it should provide, and what information will be necessary for this. To create an effective forecasting system, a significant amount of data on weather, climate, soil, vegetation, and other factors influencing the development of harmful organisms is required. Various methods and technologies, including sensors, remote sensing, and geographic information systems, can be used to collect and analyze this data. Mathematical models that consider all factors influencing the development of harmful organisms need to be developed for forecasting. These models can be based on statistical methods, machine learning, and other algorithms. After developing mathematical models, software needs to be created to implement these models and provide users with a user-friendly interface to work with the system. Before system implementation, testing and adjustment are necessary to ensure its effectiveness and the accuracy of forecasts. Errors and inaccuracies may be identified during testing, which need to be addressed. After system implementation, users must be trained to work with it and utilize the information obtained to make decisions. This is a crucial stage that helps use the system most efficiently and achieve maximum profitability in agricultural production [19].

To create an effective information system for plant protection, it is essential to combine the knowledge and expertise of specialists in plant protection and information technologies. Such systems should contain a vast amount of information about harmful organisms, methods of combating them, as well as various plant varieties and agricultural crops. Plant protection specialists can provide information on the latest trends and methods of combating harmful organisms, as well as the challenges faced by farmers. Information technology specialists can, in turn, offer the most efficient methods of data collection, processing, and storage, as well as develop relevant algorithms and software. Therefore, collaboration between plant protection and ICT specialists is crucial for creating effective information systems for plant protection [16].

2. Material and methods

Equation (1) expresses the dependency of population growth rate on the concentration of the food base, while equation (2) describes the change in the concentration of the food base depending on the population size [20].

$$W\frac{dn}{d\varsigma} = \frac{d}{d\varsigma}\left(\frac{1}{r}\frac{dn}{d\varsigma}\right) + n \tag{1}$$

$$W\frac{dr}{d\varsigma} = -\epsilon n \tag{2}$$

To find the zero isoclines, we need to set each equation to zero and solve it for the other variable. For equation (1), we obtain [21]: $\frac{dn}{d\varsigma} = 0$, at $r = 0, n = \frac{r_{-} - r}{\epsilon}$ and $\frac{dn}{d\varsigma} = 0$, at $n = 0$, On this curve, there are two special points, α(0,0) and β(0, r_). Thus, the system (1), (2) is autonomous and has zero isoclines in the phase plane (n, r), which describe situations of the absence of population and the absence of the food base [22].

Based on formula (2), it can be established that position β has the property of an unstable node if the value $(0,5r_{-}w)^2 - r_{-}$ is positive, and in the case of a negative value $(0,5r_{-}w)^2 - r_{-}$, position β will have the property of an unstable focus [23].

Numerical studies demonstrate that the results obtained from an approximate analytical solution can be expressed in terms of simple power-law dependences on two parameters: r_ and ϵ. In addition, an example was given of approximating the speed of propagation of a locust wave, which depends on these parameters in the form of a function [22, 23]:

$$w = 2,6r_{-}^{-1/2} \tag{3}$$

This equation determines the eigenvalue, which characterizes the properties of the wave solution when moving from the initial equilibrium position, taking into account the parameter β:

$$\lambda_2 = 0,5r_{-}^{1/2} \tag{4}$$

It must be emphasized that the presented model is not capable of exploring the reasons why a swarm of locusts makes long flights at high altitudes above the earth's surface, deprived of food or food sources in the environment [24].

To simulate the flight of gregarious locusts, a mathematical model consisting of nonlinear differential equations was created. The solution to this system of equations, which corresponds to the flight of a swarm of locusts, can be represented as a wave solution that moves through the swarm of locusts and causes periodic oscillations in their movement. Analysis of this solution can help in studying the flight dynamics of gregarious locusts and developing methods to combat this phenomenon [25].

The population dynamics of any insect depends on the interaction of the plant and animal kingdoms. Several models related to locust migration have been presented in the research area under consideration. One of the articles [22] presented a model describing the activation of locusts, and another [23] presented the connection between migration and air flows. Also in [22], a model of crawling migration of locusts was developed. To account for the complex dynamics of population sizes in space and time, population models use migration terms that depend on the current population level. These phenomena are explained by the fact that changes in population size lead to changes in conditions in the habitat, which in turn affects the migration characteristics of the population. Waves of population spread are also possible, when an increase in population size in one area leads to an increase in neighboring areas. To account for population waves and changes in migration intensity and direction with abundance, abundance-dependent gradient terms are used in the reaction-diffusion type equation in population models. These models are studied in a number of scientific works [3, 25–27]. Thus, migration plays an important role in population dynamics, and taking into account its effects is necessary for correct modeling of such systems:

$$\begin{cases} \dfrac{\partial u_1}{\partial t} = \dfrac{\partial}{\partial x}\left(D_1 \dfrac{\partial u_1}{\partial x}\right) + a_1 \dfrac{\partial u_1}{\partial x} + f_1(u_1, u_2), \\ \dfrac{\partial u_2}{\partial t} = \dfrac{\partial}{\partial x}\left(D_2 \dfrac{\partial u_2}{\partial x}\right) + a_2 \dfrac{\partial u_2}{\partial x} + f_2(u_1, u_2), \end{cases}$$

$$u_1|_{t=0} = u_{10}(x) \quad u_2|_{t=0} = u_{20}(x)$$

If in the model considered in [4, 5] we select some parameter values $a_i = const$, $i = 1, 2$, then the equation can be reduced to a standard reaction-diffusion type equation by replacing the independent variables $t = t,\ \xi = x + a_i t$.

$$\begin{cases} \dfrac{\partial u_1}{\partial t} = \dfrac{\partial}{\partial \xi}\left(D_1 \dfrac{\partial u_1}{\partial \xi}\right) + f_1(u_1, u_2), \\ \dfrac{\partial u_2}{\partial t} = \dfrac{\partial}{\partial \xi}\left(D_2 \dfrac{\partial u_2}{\partial \xi}\right) + f_2(u_1, u_2). \end{cases}$$

The dynamics of a biological population and its food supply are described by a mathematical model that takes into account the interaction between them. This model is formulated as a system of differential equations, where the variables represent population size and food density in time and space. These equations include many parameters, such as the rate of population growth, the level of natural mortality, the rate of reproduction, the availability of food supply in a one-dimensional model that takes into account the discreteness of the flight process of gregarious locusts, which can be expressed by a system of non-stationary equations [4–6]:

$$\begin{cases} \dfrac{\partial u_1}{\partial t} = \dfrac{\partial}{\partial x}\left(D_1 u_1^{m_1-1}\left|\dfrac{\partial u_1}{\partial x}\right|^{p_1-2}\dfrac{\partial u_1}{\partial x}\right) + a_1\dfrac{\partial u_1}{\partial x} + k_1(t)f_1(u_1, u_2) \\ \dfrac{\partial u_2}{\partial t} = \dfrac{\partial}{\partial x}\left(D_2 u_2^{m_2-1}\left|\dfrac{\partial u_2}{\partial x}\right|^{p_2-2}\dfrac{\partial u_2}{\partial x}\right) + a_2\dfrac{\partial u_2}{\partial x} + k_2(t)f_2(u_1, u_2), \end{cases} \tag{5}$$

$$x \in R_+^1, t \in R_+^1, u_1|_{t=0} = u_{10}(x), u_2|_{t=0} = u_{20}(x), a = a(t). \tag{6}$$

In this equations $D_1 u_1^{m_1-1}\left|\frac{\partial u_1}{\partial x}\right|^{p_1-2}$ and $D_2 u_2^{m_2-1}\left|\frac{\partial u_2}{\partial x}\right|^{p_2-2}$ are the diffusion coefficient, $m_1, m_2, p_1, p_2, \beta_1, \beta_2$ – positive real numbers, $u_1 = u_1(t, x) \geq 0$, $u_2 = u_2(t, x) \geq 0$ – required solutions.

Equation (5) is an extension of the basic diffusion model used to describe population growth using a logistic model [1, 2] based on the Malthus model ($f_1(u_1, u_2) = u_1$, $f_1(u_1, u_2) = u_2$, $f_2(u_1, u_2) = u_1$, $f_2(u_1, u_2) = u_2$), on the Förshulst type model ($f_1(u_1, u_2) = u_1(1 - u_2)$, $f_1(u_1, u_2) = u_2(1 - u_1)$, $f_2(u_1, u_2) = u_1(1 - u_2)$, $f_2(u_1, u_2) = u_2(1 - u_1)$), and on the Ollie type model ($f_1(u_1, u_2) = u_1(1 - u_2^{\beta_1})$, $f_1(u_1, u_2) = u_2(1 - u_1^{\beta_2})$, $f_2(u_1, u_2) = u_1(1 - u_2^{\beta_1})$, $f_2(u_1, u_2) = u_2(1 - u_1^{\beta_2})$, $\beta_1 > 1,\ \beta_2 > 1$) for the case of double nonlinear diffusion. These equations represent an optimal combination of nonlinear diffusion equations in the form ($m_i + p_i - 3 > 0$, $i = 1, 2$) fast diffusion ($2 < m_i + p_i < 3$, $i = 1, 2$), very fast diffusion ($m_i + p_i < 2$, $i = 1, 2$). Case $m_i + p_i - 3 = 0$, $i = 1, 2$ called a critical case.

To analyze the solution to equations (5) and (6), as well as to set the initial and boundary conditions, it is necessary to perform an analytical solution. For the purpose of further analysis, we present the following theorem.

Theorem 1. Let $0 \leq u_{i0} \leq 1$, $x \in R$, $i = 1, 2$ and the following conditions are met $u_i(0, x) \leq z_i(0, x)$, where $z_i(t, x) = \overline{u}_i(t) \cdot \overline{f}_i(\xi)$,

$\overline{u}_i(t), i = 1, 2$ is a solution to the equation $\begin{cases} \dfrac{d\overline{u}_1}{dt} = k_1(t)\overline{u}_1 \cdot \left(1 - \overline{u}_2^{\beta_1}\right), \\ \dfrac{d\overline{u}_2}{dt} = k_2(t)\overline{u}_2 \cdot \left(1 - \overline{u}_1^{\beta_2}\right), \end{cases}$

$\overline{f}_i(\xi) = \left(a_i - b_i \xi^{\frac{p_i}{p_i-1}}\right)_+^{\frac{p_i-1}{p_i+m_i-3}}$, where $\xi = \dfrac{x - \int_0^t a_1(t)dt}{(T+t)^{\frac{1}{p}}}$, $a_i > 0$, $b_i = \dfrac{p_i+m_i-3}{p_i^{\frac{p_i}{p_i-1}}}$, here $a_+ = \begin{cases} a, \text{ if } > 0 \\ 0, \text{ if } \le 0 \end{cases}$, $\tau_1(t) = \int [\overline{u}(t)]^{p_1+m_1-3} dt$.

Then problem (5), (6) is globally solvable. Moreover, there are estimates from two sides for solving the area $Q = \{(t,\ x);\ t > 0,\ x \in R\}$

$$\overline{u}_i(t)(T+t)^{-\gamma_i} e^{-\xi^{p_i}/4D_i} \le u_i(t,x) \le e^{\int_0^t k_i(t)dt} (T+t)^{-\gamma_i} e^{-\xi^{p_i}/4D_i},$$

where are the functions $\overline{u}_i(t)$ were previously defined.

Proof. To prove the theorem, we transform equation (5) by changing variables using the nonlinear splitting method [3], and then obtain an upper bound.

Replacement in (5) $t = t,\ \xi = x + a_i t$ and

$$u_1(t,\xi) = e^{k_1 t} v_1(\tau, x),$$
$$u_2(t,\xi) = e^{k_2 t} v_2(\tau, x)$$

will lead (5) to the form:

$$\begin{cases} \dfrac{\partial v_1}{\partial \tau} = \dfrac{\partial}{\partial x}\left(D_1 v_1^{m_1-1} \left|\dfrac{\partial v_1}{\partial x}\right|^{p_1-2} \dfrac{\partial v_1}{\partial x}\right) - k_1 e^{((\beta_1 k_2 + k_1) - (p_1+m_1-2)k_1)t} v_1 v_2^{\beta_1}, \\ \dfrac{\partial v_2}{\partial \tau} = \dfrac{\partial}{\partial x}\left(D_2 v_2^{m_2-1} \left|\dfrac{\partial v_2}{\partial x}\right|^{p_2-2} \dfrac{\partial v_2}{\partial x}\right) - k_2 e^{((\beta_2 k_1 + k_2) - (p_2+m_2-2)k_2)t} v_1^{\beta_2} v_2. \end{cases} \tag{7}$$

$v_1|_{t=0} = v_{10}(x)$, $v_2|_{t=0} = v_{20}(x)$,

By choosing $(p_1 + m_1 - 3)k_1 = (p_2 + m_2 - 3)k_2$ in equation (7), we get:

$$\begin{cases} \dfrac{\partial v_1}{\partial \tau} = \dfrac{\partial}{\partial x}\left(D_1 v_1^{m_1-1} \left|\dfrac{\partial v_1}{\partial x}\right|^{p_1-2} \dfrac{\partial v_1}{\partial x}\right) - a_1 \tau^{b_1} v_1 v_2^{\beta_1}, \\ \dfrac{\partial v_2}{\partial \tau} = \dfrac{\partial}{\partial x}\left(D_2 v_2^{m_2-1} \left|\dfrac{\partial v_2}{\partial x}\right|^{p_2-2} \dfrac{\partial v_2}{\partial x}\right) - a_2 \tau^{b_2} v_1^{\beta_2} v_2, \end{cases} \tag{8}$$

where

$$a_1 = ((p_1 + m_1 - 3)k_1)^{b_1}, a_2 = ((p_2 + m_2 - 3)k)^{b_2}$$

$$b_1 = [(\beta_1 k_2 + k_1) - (p_1 + m_1 - 2)k_1]/(p_1 + m_1 - 3)k_1,$$

$$b_2 = [(\beta_2 k_1 + k_2) - (p_2 + m_2 - 2)k_2)k_2]/(p_2 + m_2 - 3)k_2)k_2.$$

If $b_i = 0, i = 1, 2$, i.e. $\beta_1 k_2 = (p_1 + m_1 - 2)k_1$ and $\beta_2 k_1 = \sigma_2 k_2$, then:

$$\begin{cases} L_1 v_1 \equiv -\dfrac{\partial v_1}{\partial \tau} + \dfrac{\partial}{\partial x}\left(D_1 v_1^{m_1-1}\left|\dfrac{\partial v_1}{\partial x}\right|^{p_1-2}\dfrac{\partial v_1}{\partial x}\right) - a_1 v_1 v_2^{\beta_1} = 0, \\ L_2 v_2 \equiv -\dfrac{\partial v_2}{\partial \tau} + \dfrac{\partial}{\partial x}\left(D_2 v_2^{m_2-1}\left|\dfrac{\partial v_2}{\partial x}\right|^{p_2-2}\dfrac{\partial v_2}{\partial x}\right) - a_2 v_1^{\beta_2} v_2 = 0. \end{cases}$$

Function $\overline{f}_i(\xi) = e^{\frac{-\xi^p}{4D_i}}$ satisfies the equation

$$\begin{cases} \dfrac{d}{d\xi}\left(D_1 \overline{f}_1^{m_i-1}\left|\dfrac{d\overline{f}_1}{d\xi}\right|^{p_1-2}\dfrac{d\overline{f}_1}{d\xi}\right) + c\dfrac{d\overline{f}_1}{d\xi} + \gamma_1 \overline{f}_1 = 0, \\ \dfrac{d}{d\xi}\left(D_2 \overline{f}_2^{m_i-1}\left|\dfrac{d\overline{f}_2}{d\xi}\right|^{p_2-2}\dfrac{d\overline{f}_2}{d\xi}\right) + c\dfrac{d\overline{f}_2}{d\xi} + \gamma_2 \overline{f}_2 = 0. \end{cases}$$

Therefore, it can be argued that the function $(T+t)^{-\gamma_i} e^{\frac{-x^{p_i}}{4D_i(T+t)}}$ represents the upper solution of equation (8), since its gradient is always directed upward and it satisfies the initial and boundary conditions

$$\begin{cases} L\overline{f}_1 = \dfrac{d}{d\xi}\left(D_1 \overline{f}_1^{m_1-1}\left|\dfrac{d\overline{f}_1}{d\xi}\right|^{p_1-2}\dfrac{d\overline{f}_1}{d\xi}\right) + c\dfrac{d\overline{f}_1}{d\xi} + \gamma_1 \overline{f}_1 - a_1 (T+t)^{1-\gamma_1} e^{\frac{-x^{p_1}(\beta_1+1)}{4D_1(T+t)^{(\beta_1+1)}}} \le 0, \\ L\overline{f}_2 = \dfrac{d}{d\xi}\left(D_2 \overline{f}_2^{m_2-1}\left|\dfrac{d\overline{f}_2}{d\xi}\right|^{p_2-2}\dfrac{d\overline{f}_2}{d\xi}\right) + c\dfrac{d\overline{f}_2}{d\xi} + \gamma_2 \overline{f}_2 - a_2 (T+t)^{1-\gamma_2} e^{\frac{-x^{p_2}(\beta_2+1)}{4D_2(T+t)^{(\beta_2+1)}}} \le 0 \end{cases}$$

in Q at any constant T>0.

According to the solution comparison theorem [3], for solving problem (5), (6) the upper bound is true

$$\begin{cases} u_1(t,x) \le e^{k_1 t}(T+t)^{-\gamma_1} e^{\frac{-x^{p_1}}{4D_1(T+t)}}, \\ u_2(t,x) \le e^{k_2 t}(T+t)^{-\gamma_2} e^{\frac{-x^{p_2}}{4D_2(T+t)}} \end{cases}$$

in $Q = \{(t,x) : t > 0, x \in R\}$, if $\begin{cases} u_1(0,x) \le T^{-\gamma_1} e^{\frac{-x^{p_1}}{4D_1 T}}, \\ u_2(0,x) \le T^{-\gamma_2} e^{\frac{-x^{p_2}}{4D_2 T}}. \end{cases}$

Therefore, the function $e^{k_i t}(T+t)^{-\gamma_i} e^{\frac{-x^{p_i}}{4D_i(T+t)}}$ can serve as an upper bound for solving problem (5), (6).

To estimate the lower solution, we apply the nonlinear splitting method [7, 8]. Let's start by finding a solution to a system of ordinary differential equations

$$\begin{cases} \dfrac{d\overline{v}_1}{d\tau} = -a_1\overline{v}_1\overline{v}_2^{\beta_1} \\ \dfrac{d\overline{v}_2}{d\tau} = -a_2\overline{v}_1^{\beta_2}\overline{v}_2 \end{cases}$$

in the form $\overline{v}_1(\tau) = c_1(\tau + T_0)^{-\gamma_1}$, $\overline{v}_2(\tau) = c_2(\tau + T_0)^{-\gamma_2}$, $T_0 > 0$, where $c_1 = 1$, $\gamma_1 = \frac{1}{\beta_2}$, $c_2 = 1$, $\gamma_2 = \frac{1}{\beta_1}$.

Next we will obtain an estimate for the lower solution. In order to solve the system of ordinary differential equations (6) and (7), we first apply the nonlinear splitting method as described in [4]. Then we find a solution to this system of equations in the form:

$$v_1(t, x) = \overline{v}_1(t)w_1(\tau, x) \tag{9}$$

$v_2(t, x) = \overline{v}_2(t)w_2(\tau, x)$,
$\tau = \tau(t)$ is chosen like this

$$\tau_1(\tau) = \int_0^\tau \overline{v}_1^{p_1+m_1-3}(\tau)dt = \begin{cases} \frac{1}{1-\gamma_1(p_1+m_1-3)}(T+\tau)^{1-\gamma_1(p_1+m_1-3)1} & \mathit{if}\ u1 - \gamma_1(p_1+m_1-3) \neq 0, \\ \ln(T+\tau), & \mathit{if}\ 1 - \gamma_1(p_1+m_1-3) = 0. \end{cases}$$

Then when choosing $w_i(\tau, x)$, $i = 1, 2$ in equation (7) we obtain a system of equations:

$$\begin{cases} \dfrac{\partial w_1}{\partial \tau} = \dfrac{\partial}{\partial x}\left(D_1 w_1^{m_1-1}\left|\dfrac{\partial w_1}{\partial x}\right|^{p_1-2}\dfrac{\partial w_1}{\partial x}\right) + \psi_1(w_1 w_2^{\beta_1} - w_1), \\ \dfrac{\partial w_2}{\partial \tau} = \dfrac{\partial}{\partial x}\left(D_2 v_2^{m_2-1}\left|\dfrac{\partial v_2}{\partial x}\right|^{p_2-2}\dfrac{\partial v_2}{\partial x}\right) + \psi_2(w_2 w_1^{\beta_2} - w_2), \end{cases} \tag{10}$$

where

$$\psi_i = \begin{cases} \frac{\gamma_i}{(1-\gamma_1(p_1+m_1-3))\tau} & \mathit{if}\ u1 - \gamma_1(p_1+m_1-3) > 0, \\ \gamma_i c_1^{-(p_1+m_1-3)}, & \mathit{if}\ 1 - \gamma_1(p_1+m_1-3) = 0. \end{cases} \tag{11}$$

Representation of system (5) in the form (9) allows us to assume that, when $\tau \to \infty$, $\psi_i \to 0$ and

$$\begin{cases} \dfrac{\partial w_1}{\partial \tau} = \dfrac{\partial}{\partial x}\left(D_1 w_1^{m_1-1}\left|\dfrac{\partial w_1}{\partial x}\right|^{p_1-2}\dfrac{\partial w_1}{\partial x}\right), \\ \dfrac{\partial w_2}{\partial \tau} = \dfrac{\partial}{\partial x}\left(D_2 v_2^{m_2-1}\left|\dfrac{\partial v_2}{\partial x}\right|^{p_2-2}\dfrac{\partial v_2}{\partial x}\right). \end{cases} \tag{12}$$

Consequently, the solution to the system of equations (5) with initial and boundary conditions (9) will approach the solution (12).

Let $\gamma_1(p_1+m_1-3)>1$, $\gamma_1(p_1+m_1-3)=\gamma_2(p_2+m_2-3)$, $(p_2+m_2-3)(b_2+1)+\beta_2(b_1+1)=\gamma_1(p_1+m_1-3)(b_1+1)+\beta_1(b_2+1)$, $c_i>0$. Assuming in (12)$w_i(\tau(t),x)=y_i(\xi)$, $\xi=|x|/\tau_1^{1/p}$, $i=1,2$. Considering that the equation for $w_i(\tau,x)$, which does not contain lower terms, always has a self-similar solution in the case when the condition is satisfied $1-\gamma_1(p_1+m_1-3)\neq 0$,, then we can get a system of equations:

$$\begin{cases} \frac{d}{d\xi}(y_1^{m_1-1}\left|\frac{dy_1}{d\xi}\right|^{p_1-2}\frac{dy_1}{d\xi})+c\frac{dy_1}{d\xi}+\mu_1(y_1-y_1y_2^{\beta_1})=0, \\ \frac{d}{d\xi}(y_2^{m_2-1}\left|\frac{dy_2}{d\xi}\right|^{p_2-2}\frac{dy_2}{d\xi})+c\frac{dy_2}{d\xi}+\mu_2(y_2-y_2y_1^{\beta_2})=0, \end{cases} \tag{13}$$

where $\mu_i=\frac{1}{(p_i+m_i-3)}$.

Then it is easy to calculate that when $\overline{f}_i(\xi)=e^{\frac{-\xi^p}{4D_i}}$

$$\begin{cases} L\overline{f}_1=a_1\overline{v}_1^{\beta_1-(p_1+m_1-2)}\tau(y_1-y_1y_2^{\beta_1})\geq 0, \\ L\overline{f}_2=a_2\overline{v}_2^{\beta_2-(p_2+m_2-2)}\tau(y_2-y_2y_1^{\beta_2})\geq 0, \end{cases}$$

in $Q=\{(t,x):t>0,x\in R\}$ if $\begin{cases} u_1(0,x)\leq T^{-\gamma_1}e^{\frac{-x^{p_1}}{4D_1T}}, \\ u_2(0,x)\leq T^{-\gamma_2}e^{\frac{-x^{p_2}}{4D_2T}}. \end{cases} T\geq 1.$

Combining the upper and lower estimates obtained above, we can calculate an estimate for solving this problem (5), (10):

$$\left(a_i-b_i\xi^{\frac{p_i}{p_i-1}}\right)_+^{\frac{p_i-1}{p_i+m_i-3}}(T+t)^{-\gamma_i}e^{\frac{-x^{p_i}}{4D_i(T+t)}}\leq u_i(t,x)\leq e^{k_it}(T+t)^{-\gamma_i}e^{\frac{-x^{p_i}}{4D_i(T+t)}}.$$

To determine the speed of a wave $\frac{dx}{dt}$ from the last expression from the condition $e^{k_it-x^{p_i}/4D_i(T+t)}=(T+t)^{\gamma_i}$ have:

$$x(t)=(4D_ik_it(t+T)^{\frac{1}{p_i}}-\gamma_i\ln(T+t),\ i=1,2 \tag{14}$$

$$\frac{dx}{dt}=\frac{2}{p_i}(4D_ik_i)^{\frac{1}{p_i}}(t(1+T/t)^{1/2})^{\frac{2}{p_i}-1}(1+T/t)^{1/2}-(T/2)\frac{1}{t}(1+T/t)^{-1}-\gamma_i(T+t)^{-1}.$$

Which shows what if $p_i>2$, then $\frac{dx}{dt}<\infty$ at $t\to\infty$ and therefore the speed of the wave is finite at t>0.

To numerically solve problem (6), it is necessary to construct a uniform grid:

$$\omega_h=\left\{x_i=ih,\ h>0,\ i=0,1,...,n,\ hn=l\right\},$$

and time grid $\omega_{h_1}=\left\{t_j=jh_1,\ h_1>0,\ j=0,1,\ldots,n,\ \tau m=T\right\}$.

Let us replace the system of equations (7) with an implicit difference scheme and obtain a difference problem with an error $O\left(h^2+h_1\right)$.

When solving nonlinear problems numerically, one of the main problems is the choice of the correct initial approximation and the use of methods for linearizing the system of equations. To overcome this problem, we will consider the following functions:

$$v_{10}(t,x) = v_1(t) \cdot \left(a_1 - b_1 \xi^{\frac{p_1}{p_1-1}}\right)_+^{\frac{p_1-1}{p_1+m_1-3}}, \quad v_{20}(t,x) = v_2(t) \cdot \left(a_2 - b_2 \xi^{\frac{p_2}{p_2-1}}\right)_+^{\frac{p_2-1}{p_2+m_2-3}},$$

where $v_1(t) = e^{kt}\overline{v}_1(t)$ and $v_2(t) = e^{kt}\overline{v}_2(t)$ functions defined above.

Record $(a)_+$ means $(a)_+ = \max(0, a)$. The functions described above have the property of a finite speed of propagation of disturbances [4, 7]. Since these functions $v_{i0}(t,x), i = 1, 2$ have the property of a finite speed of propagation of disturbances [4, 7], then they can be used in the numerical solution of problem (7) with $\beta_1 > p_1 + m_1 - 3$ as an initial approximation.

2 Results

Below are the results of numerical experiments for different parameter values (see Figs. 3, 4 and 5).

Parameter values	Results of a computational experiment
$\sigma = 1.1, p = 2.1$ $k_1 = 0.1$ $k_2 = 0.25$ $eps = 10^{-3}$	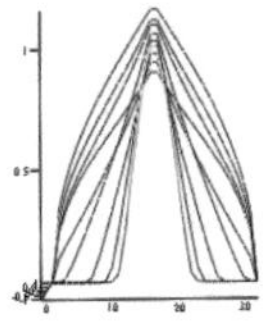
$\sigma = 1.3,\ p = 3$ $k_1 = 5$ $k_2 = 7$ $eps = 10^{-3}$	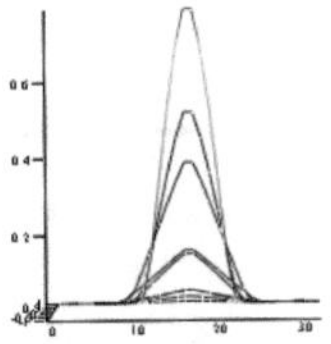

Fig. 3. Dynamics of the Malthusian population.

During this study, an analysis of the ecology and fauna of locusts in the region was carried out, as well as modeling the state of the population of gregarious locusts. The work aims to gain an in-depth understanding of the relationships in the biological system with the aim of developing numerical schemes and methods for effectively managing and predicting the dynamics of locust populations. One of the main aspects of the study is to analyze the characteristics of the ecology and fauna of locusts in the region. The findings provide a better understanding of the factors influencing the life cycle and behavior of locusts, an important step for developing effective control strategies. Studying the properties of mathematical models of a nonlinear population of gregarious locusts contributes to the development of more accurate and realistic models that reflect complex

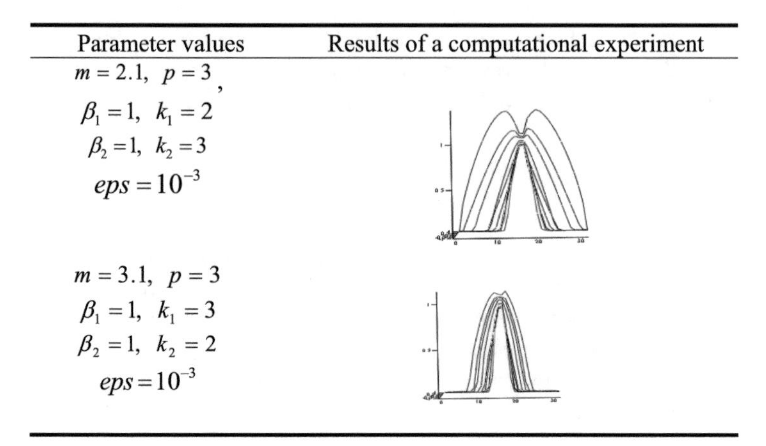

Parameter values	Results of a computational experiment
$m = 2.1,\ p = 3$, $\beta_1 = 1,\ k_1 = 2$, $\beta_2 = 1,\ k_2 = 3$, $eps = 10^{-3}$	
$m = 3.1,\ p = 3$, $\beta_1 = 1,\ k_1 = 3$, $\beta_2 = 1,\ k_2 = 2$, $eps = 10^{-3}$	

Fig. 4. Logistics population dynamics.

Parameter values	Results of a computational experiment
$m = 3.1,\ p = 3$, $\beta_1 = 7,\ k_1 = 7$, $\beta_2 = 9,\ k_2 = 5$, $eps = 10^{-3}$	
$m = 1.3,\ p = 3$, $\beta_1 = 3,\ k_1 = 5$, $\beta_2 = 7,\ k_2 = 3$, $eps = 10^{-3}$	

Fig. 5. Dynamics of the Ollie effect.

interactions within the population. This makes it possible to more accurately predict population dynamics and more effectively plan control measures. The use of computer modeling of the processes of multicomponent cross-diffusion systems of the biological population of gregarious locusts based on the principles of comparison of solutions makes it possible to take into account complex interactions between different parts of the population and the environment. This leads to more accurate and realistic models and allows for more accurate predictions of long-term trends in locust populations. One of the key stages of the study was the search for suitable initial approximations, including using the linearization method. This method is an important tool for facilitating calculations and increasing the stability of numerical algorithms. The developed algorithms and programs will serve as the basis for conducting a computational experiment. This stage of the study not only allows us to verify the correctness and effectiveness of the results obtained, but also serves as the basis for future applications and practical solutions in the field of locust population management. Visualization of the solution plays an important role in presenting the results and assessing their quality. A clear representation of locust population dynamics using visualization allows you to better understand the modeling results and make informed conclusions about the state of the population. Overall, this study not only expands our knowledge of locust ecology and fauna, but also provides tools for more effective management and control of locust populations in different ecosystems.

3 Conclusion

Research into the ecology and fauna of locusts, as well as mathematical modeling of nonlinear gregarious locust populations, represents a significant step towards understanding and managing these populations in different ecosystems. The results and conclusions of this study highlight the need for in-depth analysis of locust population dynamics to develop effective control strategies.

The use of computer models to analyze cross-diffusion processes in multicomponent biological population systems of gregarious locusts, based on nonlinear splitting and solution comparison principles, as well as convective transport modeling, provide improved insight into the dynamics of locust populations. These methods are important for developing management strategies as well as for more accurately predicting changes in populations.

The study of nonlinear processes of a multicomponent cross-diffusion system of a biological population of gregarious locusts allows us to identify new effects associated with nonlinearity, which in turn enriches our understanding of the relationships within the population. This new knowledge may be key to the formation of more accurate and realistic models that are superior in predictive ability to linear models.

The development of algorithms and software systems for solving problems of modeling and managing locust populations provides the opportunity to automate the process and conduct a computational experiment. This greatly simplifies the study of different locust management scenarios and also improves the efficiency of developing control strategies.

Overall, research in this area contributes to increasing our knowledge of the ecology and fauna of locusts, and also provides concrete tools for developing effective strategies for managing their populations in different ecosystems. These scientific findings have practical implications and can serve as a basis for developing more effective locust management programs in the future.

References

1. Marie, J.: Nonlinear Diffusion Equations in Biology. Mir, Moscow (1983)
2. Kolmogorov, A.N., Petrovsky, I.G., Piskunov, N.S.: Study of the equation of diffusion coupled with an increase in the amount of substance and its application to a biological problem. Bull. Moscow State Univ. Math. Mech. **1**, 1–25 (1937)

3. Aripov, M.: Method of Standard Equations for Solving Nonlinear Boundary Value Problems. Tashkent, Fan (1988)
4. Belotelov, N.V., Lobanov, A.I.: Population models with nonlinear diffusion. Math. Model. **12**, 43–56 (1997)
5. Samarsky, A.A., Galaktionov, V.A., Kurdyumov, S.P., Mikhailov, A.S.: Modes with Sharpening in Problems for Quasilinear Parabolic Equations. Moscow, Science (1987)
6. Aripov, M.M., Muhamediyeva, D.K.: On the properties of the solutions of the problem of cross-diffusion with the dual nonlinearity and the convective transfer. J. Phys.: Conf. Ser. **1441**, 012131 (2020). https://doi.org/10.1088/1742-6596/1441/1/012131
7. Muhamediyeva, D.K.: Study parabolic type diffusion equations with double nonlinearity. J. Phys.: Conf. Ser. **1441**, 012151 (2020). https://doi.org/10.1088/1742-6596/1441/1/012151
8. Muhamediyeva, D.K.: Two-dimensional Model of the Reaction-Diffusion with Nonlocal Interaction. In: 2019 International Conference on Information Science and Communications Technologies (ICISCT), pp. 1–5, Tashkent, Uzbekistan (2019). https://doi.org/10.1109/ICISCT47635.2019.9011854
9. Gapparov, F.A., Lachininsky, A.V., Sergeyev, M.G.: Outbreaks of the Moroccan locust in Central Asia. Plant Protect. Quarantine **3**, 22–24 (2008)
10. Iskak, S., Agibaev, A.Zh., Taranov, B.T., Kalmakbayev, T.Zh., Kambulin, V.E.: Spread of migratory locusts and protective measures against them in Kazakhstan. In: Proceedings of the National Academy of Sciences of the Republic of Kazakhstan. Series of Agricultural Sciences, vol. 5, pp. 11–20 (2012).
11. Kurishbayev, A.K., Azhbenov, V.K.: Preventive approach to solving the locust invasion problem in Kazakhstan and border areas. Kazakh Agrotechnical University named after S. Seifullin. Sci. Bull. **1**(76), 42–52 (2013)
12. Azhbenov, V.K., Kostyuchenkov, N.V., Sarbaev, A.T., Baybusenov, K.S., Suleimenova, Z.Sh., Zagaynov, N.A.: Italian locust (Calliptamus italicus L.) in Kazakhstan. Astana (2017)
13. Kovalenkov, V.G., Kuznetsova, O.V.: How to contain the spread of the Italian locust. Plant Protect. Quarantine **9**, 14–17 (2011)
14. Stamo, P.D., Kovalenkov, V.G., Kuznetsova, O.V., Nikitenko, Y.: The Moroccan locust again in the Stavropol region. Plant Protect. Quarantine **2**, 14–20 (2013)
15. Long, Z., Lecoq, M., Latchininsky, A.V., Hunter, D.: Locust and grasshopper management. Annu. Rev. Entomol. **64**, 15–34 (2019)
16. Yakhyaev, K.K., Kholmuradov, E.A.: Automation of Forecasting the Development and Spread of Pests and Diseases of Agricultural Crops. FAAKANRuz, Tashkent (2005)
17. Muhamediyeva, D.K.: The property of the problem of reaction diffusion with double nonlinearity at the given initial conditions. Int. J. Mech. Prod. Eng. Res. Dev. **9**(3), 1095–1106 (2019)
18. Baibussenov, K.S., Sarbaev, A.T., Azhbenov, V.K., Harizanova, V.B.: Environmental features of population dynamics of hazard nongregarious locusts in northern Kazakhstan. Life Sci. J. **11**(10), 277–281 (2014)
19. Baibussenov, K.S., Sarbaev, A.T., Azhbenov, V.K., Harizanova, V.B.: Predicting the phase state of the abundance dynamics of harmful non-gregarious locusts in Northern Kazakhstan and substantiation of protective measures. Biosci. Biotechnol. Res. Asia **12**(2), 1535–1543 (2015)
20. Zhizhin, G.V.: Self-regulating Waves of Chemical Reactions and Biological Populations. Nauka, St. Petersburg (2004)
21. Muhamediyeva, D.T., Niyozmatova, N.A.: Approaches to solving the problem of fuzzy parametric programming in weakly structured objects. J. Phys.: Conf. Series **1260**(10), 102011 (2019)
22. Zhizhin, G.V.: Dissipative Structures in Chemical, Geological and Environmental Systems. Nauka, St. Petersburg (2005)

23. Muhamediyeva, D.T.: Model of estimation of success of geological exploration in perspective. Int. J. Mech. Product. Eng. Res. Dev. **8**(2), 527–538 (2018)
24. Zhizhin, G.V.: Combustion Waves with Distributed Zones of Chemical Reactions: (Non-asymptotic Theory of Combustion). Werner Regen Publishing House, St. Petersburg (2008)
25. Zhizhin, G.V., Selikhovkin, A.V.: Mathematical Modeling of the Development and Spread of Populations of Insect Stem Pests in Russian Forests. Publishing House of SPbGLTU, St. Petersburg (2012)
26. Topaz, C.M., Bernoff, A.J., Logan, S., Toolson, W.: A model for rolling swarms of locusts. Eur. Phys. J. Special Topics **157**, 93–109 (2008)
27. Yakhyaev, K.K., Abdullaeva, K.Z.: Automated system for monitoring the development and spread of pests of agricultural crops. Science and World **2:5**(33), 94–96 (2016)

Designing the Architecture of a Multi-agent City Management System Using Advanced Object-Oriented Modeling

I. B. Bondarenko, V. L. Litvinov, D. A. Pelikh(✉), D. A. Rozhkova, and F. V. Filippov

The Bonch-Bruevich Saint-Petersburg State University of Telecommunication, 22 Bolshevikov Avenue, Saint Petersburg 193232, Russian Federation
pelih.da@sut.ru

Abstract. The development of digitalization in the field of public sector economy is considered. The application of multi-agent approach in solving urban planning problems is proposed. The difficulties encountered in this approach are described. The method of planning actions of intellectual information agents in conditions of stochastic nature of the environment is proposed. The main groups of agents are singled out, their tasks and options for solving arising problems are described. The interaction between the elements of different levels of the system is described. The functional of coordinating agents, including the management of the reward function, is presented. The necessity of user's interaction with the system is justified, and also it is indicated at what levels he influences it. The architecture of the multi-agent system of city management at the conceptual level, taking into account the heterogeneity of its components, is given. The functional elements of this system and interrelations between them are described. The application of extended object-oriented modeling in analyzing the functionality of the system components is substantiated. The key features of the analysis are described. Extended object-oriented models of functioning of the main groups of agents and user actions are presented. Formulas for calculating the mathematical expectation of the execution time of the sequence of actions of these groups are given. Graphs of the probability density function for the corresponding components of the system are plotted.

Keywords: Multi-Agent Systems · Architecture · Urban Resource Management · Advanced Object-Oriented Models

1 Introduction

The integration of digital services into various social areas is increasing due to the development of the digital economy. Digitalization also affects the public sector economy. Various public goods provided by government, such as lighting, heating, electrification, roads, and much more can be received only through a centralized management system. In developed countries, on average, 30 to 50 percent of GDP is redistributed through the state budget.

A. Gibadullin (Ed.): DITEM 2023, LNNS 942, pp. 106–117, 2024.
https://doi.org/10.1007/978-3-031-55349-3_9

A significant part of public goods is accumulated in the urban infrastructure, which is managed by various services which actions may not be coordinated or even be used in their own interests. Thus, the development of an information system that coordinates the actions of these services and helps to optimize the decisions they make, is a relevant city management task. This problem solution is considered on the basis of a multi-agent approach within the framework of this paper.

2 Problems of Designing Multi-agent Systems

The complexity of this approach lies in the heterogeneity of the system components and the stochasticity of the environment. In [1–3], the problems of designing and managing intelligent information agents are considered. In [4], the architecture of planning service-oriented systems under conditions of uncertainty is given. In [5], the application of a multi-agent approach to solving urban planning problems is justified. In [6–9] it is shown how this approach can be applied in solving applied problems. And [10] describes how a multi-agent system can be made more secure.

You may several a number of problems when designing multi-agent systems. For example, the goal of agents responsible for road traffic is to minimize congestion, that is, to minimize the number of cars that move at low speed along the highway. The optimal solution to this problem would be to ban the movement of vehicles along this section, which guarantees the absence of congestion, but in fact leads to even worse consequences. Another example may be the task of maximizing the capacity of the road network, that is, increasing the number of vehicles passing per unit of time. The optimal solution to this problem may be to increase the area of the roadway, and to achieve maximum results, the system can offer to demolish any buildings and parks in the city and build a road in their place.

Consequently, the system should be built in such a way that when one agent or a group of agents make decisions reflecting the actions of the city management subjects, the quality indicators of other agents functioning would not deteriorate dramatically. In addition, it is necessary to understand exactly what is taken as goals. For example, in the case of optimizing road traffic, maximizing the number of vehicles moving per unit of time will not be the best task. The reason is that a public transport traffic which carries a much larger number of people than a personal car will not be considered.

3 Description of Multi-agent City Management Architecture Development

The key stage in the development of a multi-agent city management system is the definition of its functional elements and the relationships between them, that is the design of architecture. Let's take a closer look at this question. Initially, the system receives raw data: it describes the current state of the city. This may be data on the routes of urban transport, the number and location of social facilities or the roadway condition. This data should be sorted out by dividing it into specific areas of urban regulation and grouped by common characteristics (e.g. belonging to one district). The solution of these tasks is assigned to the first group agents of data agents (agent $d_1 - d_n$).

Agents make observations and decisions based on needs which are compared with input data. By needs, we mean a set of parameters, the values of which the agent seeks to bring to ideal. The ratio of the number of passengers transported per unit of time to the number of pieces of equipment for a specific route can act as a need for agents from the field of public transport. For agents from the field of social infrastructure, this may be the ratio of the number of schools to the population density in a particular area.

In order to properly configure the system, the participation of the user is required. By user we mean a group of specialists competent in specific areas of urban regulation. The user works with the system through an interactive graphical interface. The user can set or adjust the ideal values of the agents' needs depending on the desired result. The user can also manage the influence of agents on the system. When the weight coefficients of agents change, their contribution to the problem solution changes. The result of the system is the building of a plan which reflects the predicted state of the urban infrastructure after some discrete time. It is most correct to take one year as the sampling period, since the city budget is drawn up for this time, committee plans are formed, and so on. However, during this period you can take another time. The user can choose time parameter for a plan, as well as adjust the one that is already built if necessary. For example, if the system for some reason did not consider all the restrictions and proposed to dismantle the cultural heritage object, the user can set this restriction in a strict way and the system will rebuild the plan taking it into account. This article does not address the issue of system performance, although this is extremely important because the user may want to rebuild plans after each change in the parameter of a need or agent due to long time requirement when operating with a huge amount of data.

The task of the second-level agents, the knowledge group (agent $k_1 - k_n$), is to build observations. Based on the needs, they identify and offer the next level of consideration of problematic situations. For example, if the area has a high population density and a small number of hospitals, agents from the second group recognize this as a problem. Third-level agents, solution groups (agent $a_1 - a_n$), are engaged in solving such problems. The agents themselves represent the subjects of urban regulation as a reflection of the situation in the real world: these may be city committees, municipal services, federal structures, and so on. The agent enters the results in the appropriate table after agreeing on the plan. It reflects the most optimal predicted state of the urban infrastructure after a certain time.

Figure 1 shows a conceptual diagram of the architecture of a multi-agent urban resource management system.

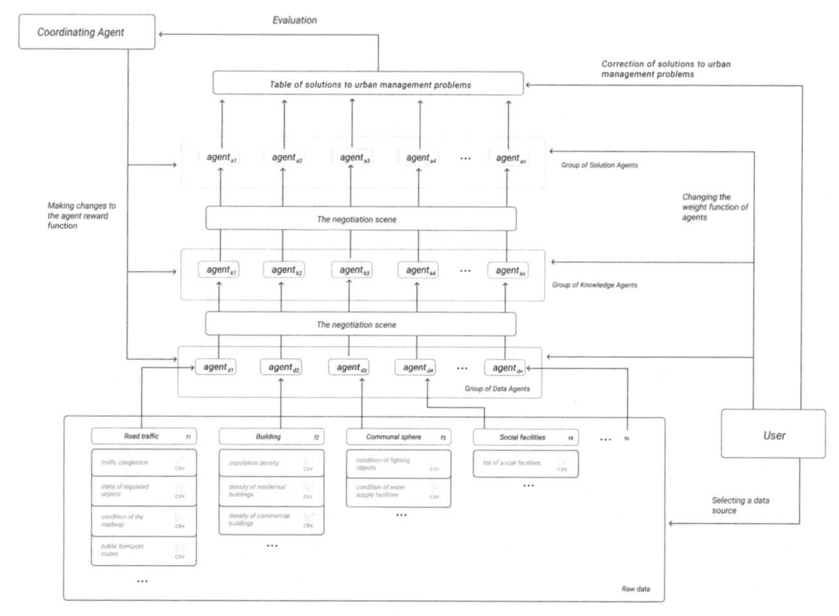

Fig. 1. Conceptual scheme of the architecture of a multi-agent urban resource management system.

4 Formation of Extended Object-oriented Models

Extended object-oriented modeling is used to describe the functionality of individual components of the above scheme. The application of this method for is described in [11]. Decision-making processes by agents have a stochastic nature, therefore, the main indicator of the quality of the functioning of the system elements is the mathematical expectation of the execution time of a certain process. The chosen methodology and quality indicators will allow for further modernization of the architecture of the designed system or the introduction of new functional elements into it to analytically assess the effectiveness of the implemented measures.

To analyze the constructed model in order to find the dependencies of mathematical expectations of the functioning time of groups of agents on their model parameters, the convolution method is chosen. When implementing the convolution method, two typical fragments are distinguished: convolution of alternative processes and convolution of the cycle. The rules for converting alternative processes and cycles are applied.

An extended object-oriented model of the functionality of data group agents is shown in Fig. 2.

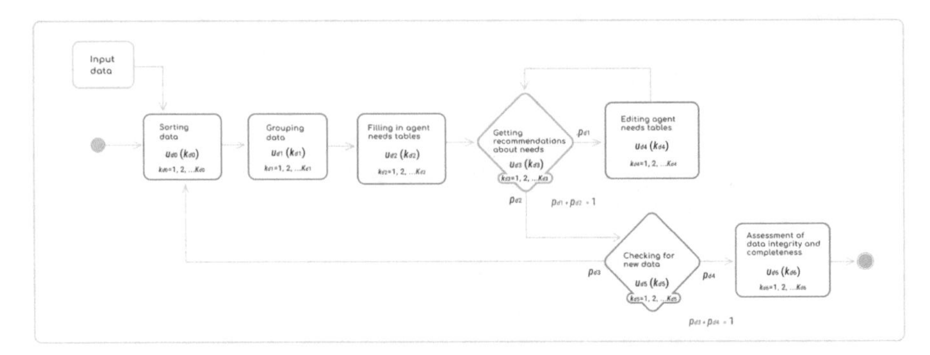

Fig. 2. An extended object-oriented model of the functionality of data group agents.

Where $u_{d0}(k_{d0})$, $u_{d1}(k_{d1})$, $u_{d2}(k_{d2})$, $u_{d3}(k_{d3})$, $u_{d4}(k_{d4})$, $u_{d5}(k_{d5})$, $u_{d6}(k_{d6})$ are the probability distribution densities of the execution time of data sorting processes, data grouping, filling in agent needs tables, getting recommendations about needs, editing agent need tables, checking for new data, assessment of data integrity and completeness accordingly.

k_{di} – discrete execution time of the *i*-th process;
K_{di} – the upper bound of the discrete execution time of the *i*-th process;

The sum of the probabilities of alternative processes (p_{d1}, p_{d2}) and (p_{d3}, p_{d4}) must be equal to one.

Mathematical expectation of the execution time of a sequence of actions of agents of a data group E $[k_{d\,0,1,2,3,4,5,6}]$:

$$\mathrm{E}[k_{d0,1,2,3,4,5,6}] = \mathrm{E}[k_{d0}] + \mathrm{E}[k_{d1}] + \mathrm{E}[k_{d2}] + p_{d2}(\mathrm{E}[k_{d3}] + \mathrm{E}[k_{d5}] + p_{d3}(\mathrm{E}[k_{d0}] + \mathrm{E} \\ ([k_{d1}] + \mathrm{E}[k_{d2}]))/(1 - p_{d3}) + p_{d1}(\mathrm{E}[k_{d4}] + p_{d1}\mathrm{E}[k_{d3}])/(1 - p_{d1}) + p_{d4}\mathrm{E}[k_{d6}]$$

An extended object-oriented model of the functionality of knowledge group agents is presented in Fig. 3.

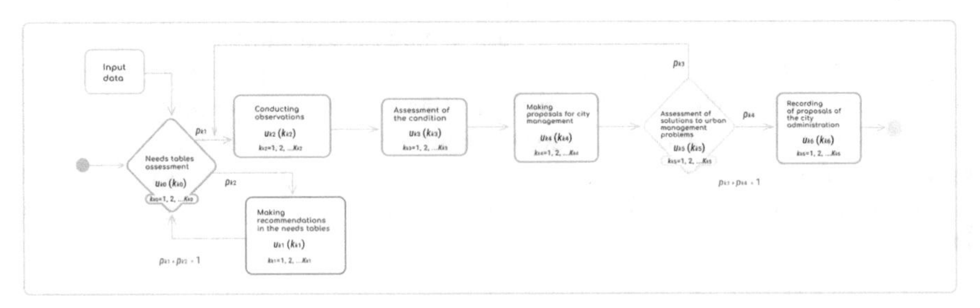

Fig. 3. Extended object-oriented model of the functionality of knowledge group agents.

Where $u_{k0}(k_{k0})$, $u_{k1}(k_{k1})$, $u_{k2}(k_{k2})$, $u_{k3}(k_{k3})$, $u_{k4}(k_{k4})$, $u_{k5}(k_{k5})$, $u_{k6}(k_{k6})$ – the probability distribution densities of the execution time of the processes of assessing the tables of needs, making recommendations to the tables of needs, conducting observations, assessing the state, making proposals for city management, evaluating solutions to urban management tasks, recording proposals of the city administration, respectively.

K_{ki} – discrete execution time of the i-th process;
K_{ki} – the upper bound of the discrete execution time of the i-th process;

The sum of the probabilities of alternative processes (p_{k1}, p_{k2}) and (p_{k3}, p_{k4}) must be equal to one.

Mathematical expectation of the execution time of the sequence of actions of the agents of the knowledge group E $[k_{k\,0,1,2,3,4,5,6}]$:

$$E[k_{k0,1,2,3,4,5,6}] = E[k_{k0}] + E[k_{k1}] + p_{k1}(E[k_{k2}] + E[k_{k3}] + E[k_{k4}]) + (E[k_{k5}] + p_{k3}p_{k1}(E[k_{k2}] + E[k_{k3}] + E[k_{k4}]))/(1 - p_{k3}) + p_{k2}(E[k_{k1}] + p_{k2}E[k_{k0}])/(1 - p_{k2}) + p_{k4}E[k_{k6}]$$

An extended object-oriented model of the functionality of the agents of the solution group is shown in the Fig. 4.

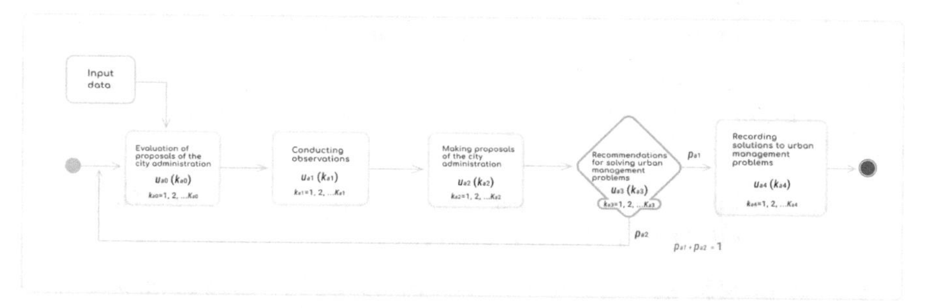

Fig. 4. Extended object-oriented model of the functionality of the agents of the solution group.

Where $u_{a0}(k_{a0})$, $u_{a1}(k_{a1})$, $u_{a2}(k_{a2})$, $u_{a3}(k_{a3})$, $u_{a4}(k_{a4})$ – the probability distribution densities of the execution time of the processes of evaluating the proposals of the city administration, conducting observations, evaluating the proposals of the city administration, offering recommendations for solving the tasks of the city administration, recording solutions to urban management problems, respectively.

k_{ai} – discrete execution time of the i-th process;
K_{ai} – the upper bound of the discrete execution time of the i-th process;

The sum of the probabilities of alternative processes (p_{a1}, p_{a2}) must be equal to one.

Mathematical expectation of the execution time of the sequence of actions of the agents of the decision group E $[k_{a\,0,1,2,3,4}]$:

$$E[k_{a0,1,2,3,4}] = E[k_{a0}] + E[k_{a1}] + E[k_{a2}] + (E[k_{a3}] + p_{a2}(E[k_{a0}] + E[k_{a1}] + E[k_{a2}]))/(1 -- p_{a2}) + p_{a1}E[k_{a4}]$$

The extended object-oriented model of coordination agents functionality is shown in the Fig. 5.

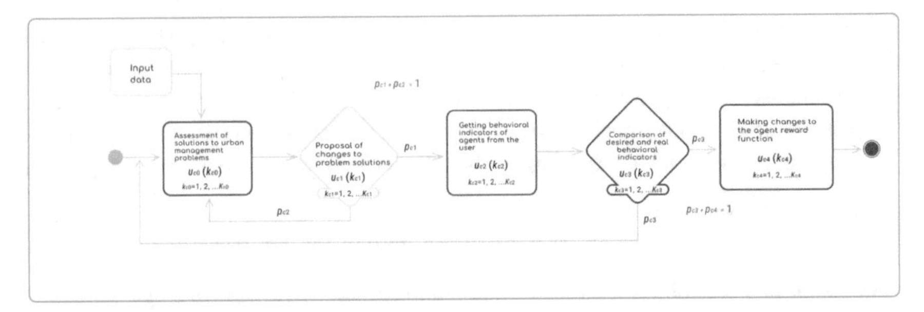

Fig. 5. Extended object-oriented model of coordination agents functionality.

Where $u_{c0}(k_{c0})$, $u_{c1}(k_{c1})$, $u_{c2}(k_{c2})$, $u_{c3}(k_{c3})$, $u_{c4}(k_{c4})$ – the probability distribution density of the execution time of the processes of assessment of solutions to urban management problems, proposal of changes to problem solution, getting behavioral indicators of agents from the user, comparing desired and real behavioral indicators, making changes to the agent's reward function accordingly.

k_{ci} – discrete execution time of the i-th process;
K_{ci} – the upper bound of the discrete execution time of the i-th process;

The sum of the probabilities of alternative processes (p_{c1}, p_{c2}) must be equal to one.

Mathematical expectation of the execution time of a sequence of actions of coordinating agents E $[k_{c\,0,1,2,3,4}]$:

$$\mathrm{E}\left[k_{c0,1,2,3,4}\right] = \mathrm{E}\left[k_{c0}\right] + (\mathrm{E}\left[k_{c1}\right] + p_{c2}\mathrm{E}\left[k_{c0}\right])/(1 - -p_{c2}) + p_{c1}\mathrm{E}\left[k_{c2}\right] + p_{c1}(\mathrm{E}\left[k_{c3}\right] + p_{c3}\mathrm{E}\left[k_{c0}\right] + p_{c3}$$
$$\mathrm{E}\left[k_{c1}\right] + p_{c3}\mathrm{E}\left[k_{c2}\right])/(1 - -p_{c3}) + p_{c4}\mathrm{E}\left[k_{c4}\right]$$

An extended object-oriented model of user actions is shown in the Fig. 6.

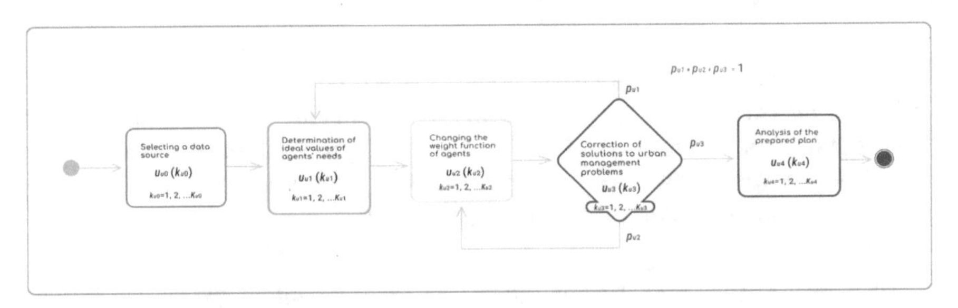

Fig. 6. Extended object-oriented model of user actions.

Where $u_{u0}(k_{u0})$, $u_{u1}(k_{u1})$, $u_{u2}(k_{u2})$, $u_{u3}(k_{u3})$, $u_{u4}(k_{u4})$ – the probability distribution density of the execution time of the processes of selecting a data source, determination of ideal values of agents' needs, changing the weight function of agents, correction of solutions to urban management problems, analysis of the prepared plan accordingly.

k_{ui} – discrete execution time of the i-th process;

K_{ui} – the upper bound of the discrete execution time of the *i*-th process;s

The sum of the probabilities of alternative processes (p_{u1}, p_{u2}) must be equal to one. Mathematical expectation of the execution time of user actions E [$k_{u\,0,1,2,3,4}$]:

$$\text{E}\,[k_{u0,1,2,3,4}]=\text{E}\,[k_{u0}]+\text{E}\,[k_{u1}]+\text{E}\,[k_{u2}]\,(1+p_{u1})+\text{E}\,[k_{u3}]\,(1+p_{u1})+(\text{E}\,[k_{u3}]+p_{u1}\text{E}\,[k_{u1}]+p_{u1}\text{E}\,[k_{u2}])/(1-p_{u1})+(\text{E}\,[k_{u3}]+p_{u2}\text{E}\,[k_{u2}])/(1--p_{u2})+p_{u3}\text{E}\,[k_{u4}]$$

5 Result of Modelling

For the purpose of visual representation of the results of extended object-oriented modeling, Markov chain probability density functions with discrete time are constructed to describe the functionality of data group agents (Fig. 7), knowledge group agents (Fig. 8), decision group agents (Fig. 9), coordination agents (Fig. 10) and user actions (Fig. 11). The general view of the matrix description of the functionality of the system elements is as follows:

$$P_d = \begin{matrix} [0, 1, 0, 0, 0, 0, 0] \\ [0, 0, 1, 0, 0, 0, 0] \\ [0, 0, 0, 1, 0, 0, 0] \\ [0, 0, 0, 0, 0.5, 0.5, 0] \\ [0, 0, 0, 1, 0, 0, 0] \\ [0.5, 0, 0, 0, 0, 0, 0.5] \\ [0, 0, 0, 0, 0, 0, 0] \end{matrix}$$

$$P_k = \begin{matrix} [0, 0.5, 0.5, 0, 0, 0, 0] \\ [1, 0, 1, 0, 0, 0, 0] \\ [0, 0, 0, 1, 0, 0, 0] \\ [0, 0, 0, 0, 1, 0, 0] \\ [0, 0, 0, 0, 0, 1, 0] \\ [0, 0, 0.5, 0, 0, 0, 0.5] \\ [0, 0, 0, 0, 0, 0, 0] \end{matrix}$$

$$P_a = \begin{matrix} [0, 1, 0, 0, 0] \\ [0, 0, 1, 0, 0] \\ [0, 0, 0, 1, 0] \\ [0.5, 0, 0, 0, 0.5] \\ [0, 0, 0, 0, 0] \end{matrix}$$

$$P_c = \begin{matrix} [0, 1, 0, 0, 0] \\ [0.5, 0, 0.5, 0, 0] \\ [0, 0, 0, 1, 0] \\ [0.5, 0, 0, 0, 0.5] \\ [0, 0, 0, 0, 0] \end{matrix}$$

$$P_u = \begin{matrix} [0, 1, 0, 0, 0] \\ [0, 0, 1, 0, 0] \\ [0, 0, 0, 1, 0] \\ [0, 0.33, 0.33, 0, 0.33] \\ [0, 0, 0, 0, 0] \end{matrix}$$

where P_d – a square matrix of transitions in a set of discrete states for the actions of data group agents, P_k – knowledge group agents, P_a – solution group agents, P_c – coordinating agents, P_u – user. In accordance with the theory of Markov chains, the probability distribution density of the execution time of a sequential action u (k) is calculated according to the Formula 1.

$$u(k) = P_{1,n}^{(k)} - P_{1,n}^{(k-1)} \tag{1}$$

The described methods and the derived analytical relations allow us to formalize the process of analyzing the functionality of the components of a multi-agent system.

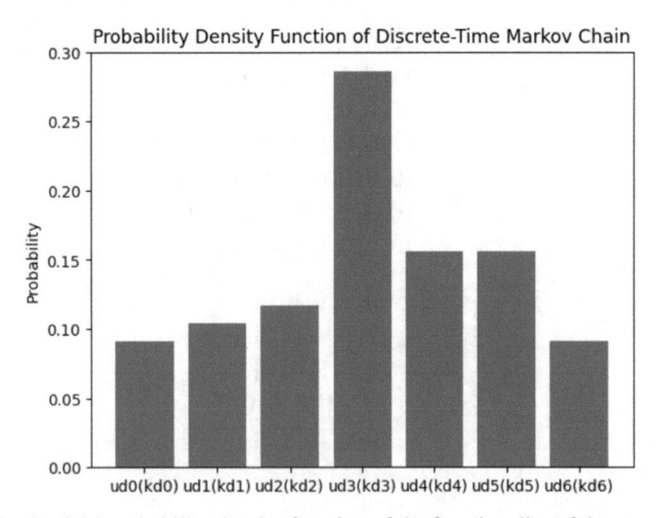

Fig. 7. Graph of the probability density function of the functionality of data group agents.

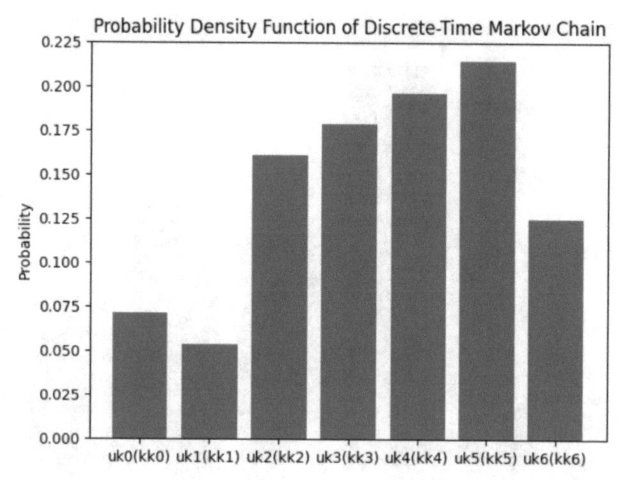

Fig. 8. Graph of the probability density function of the functionality of the agents of the knowledge group.

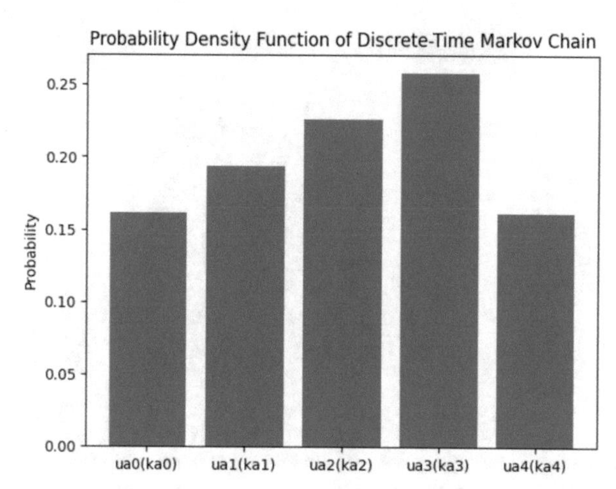

Fig. 9. Graph of the probability density function of the functionality of the agents of the solution group.

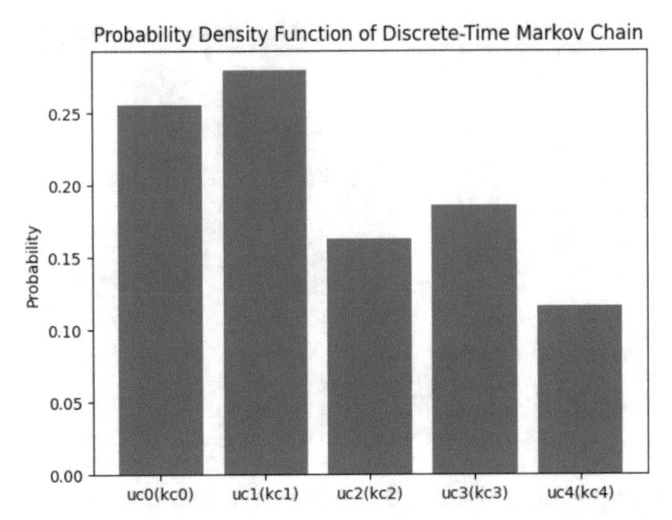

Fig. 10. Graph of the probability density function of the functionality of coordination agents.

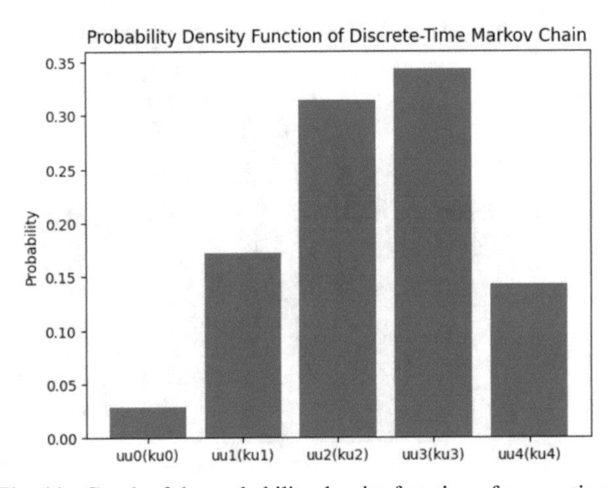

Fig. 11. Graph of the probability density function of user actions.

6 Conclusion

Digitalization in the public sector can simplify and improve the quality of management of public resources, including urban ones. The use of a multi-agent approach will allow the information system to make decisions effectively in the conditions of the stochastic nature of the environment. Therefore, the development of multi-agent city management systems is becoming an urgent task at the moment.

In order to solve this problem, the main groups of agents are identified, their tasks and solutions to emerging problems are described. The interaction between elements of different levels of the system is described. The necessity of user interaction with the system is justified. The architecture of a multi-agent city management system is presented. The use of advanced object-oriented modeling in the analysis of the functionality of system components is justified. Extended object-oriented models of the functioning of the main groups of agents and user actions are presented. Mathematical formalizations are also given, which allow analyzing the functionality of the system components. Probably, the materials given in this article can be useful in solving other applied problems using a multi-agent approach.

References

1. Ptitsyna, L.K., Shevchenco, N.E.Z., Ptitsyn, N.A., Belov, M.P.: Development of methods for modeling action planners of intelligent information agents. In: Proceedings of 2023 26th International Conference on Soft Computing and Measurements, SCM 2023, pp. 41–44 (2023)
2. Wang, C., Wang, J., Wu, P., Gao, J.: Consensus problem and formation control for heterogeneous multi-agent systems with switching topologies. Electronics **11**, 2598 (2022)
3. Salih, N.K., Zang, T., Viju, G.K., Mohamed, A.A.: Autonomic management for multi-agent systems. IJCSI Int. J. Comput. Sci. **8**(5), 338–341 (2022)
4. Alves, F., Rocha, A.M.A.C., Pereira, A.I., Leitao, P.: Distributed scheduling based on multi-agent systems and optimization methods. In: De La Prieta, F., et al. (eds.) Highlights of Practical Applications of Survivable Agents and Multi-Agent Systems. The PAAMS Collection: International Workshops of PAAMS 2019, Ávila, Spain, June 26–28, 2019, Proceedings, pp. 313–317. Springer International Publishing, Cham (2019). https://doi.org/10.1007/978-3-030-24299-2_27
5. Ptitsyna, L.K., Shevchenko, N.E.S., Belov, M.P., Ptitsyn, A.V.: Planning architecture of service-oriented systems under uncertainty. In: Proceedings of 2020 23rd International Conference on Soft Computing and Measurements, SCM 2020, pp. 101–104 (2020)
6. Pelikh, D.A., Ptitsyna, L.K.: Multi-agent systems in urban resource management. Actual problems of infotelecommunications in science and education. In: XII International Scientific-technical and Scientific-methodical Conference, vol. 4, pp. 690–694 (2023)
7. Iannino, V., Colla, V., Mocci, C., Matino, I.: Multi-agent systems to improve efficiency in steelworks. Matériaux Tech. **109**, 502 (2021)
8. Skobelev, P.O., Lakhin, O.I., Mayorov, I.V., Simonova, E.V.: Adaptive multi-agent planning of industrial resources based on ontology. Inform. Control Syst. **6**, 105–117 (2018)
9. Zhao, J., Dai, F., Song, Y.: A distributed optimal formation control for multi-agent system of UAVs. Proc. Int. Conf. Artif. Life Robot. **27**, 803–807 (2022)
10. Aquib, M., Dimitra, P.: Adversary detection and resilient control for multi-agent systems. IEEE Trans. Control Netw. Syst. **10**, 355–367 (2022)
11. Ptitsyna, L.K., Zharanova, A.O., Ptitsyn, N.A., Belov, M.P.: Extended object-oriented modeling of intelligent information agent planners. In: Proceedings of 2022 25th International Conference on Soft Computing and Measurements, SCM 2022, pp. 60–63 (2023)

Analysis of the Impact of Fintech on the Economy: Japan and Russia Case

Yevgeniya Medvedkina(✉) and Olga Borisenko

Don State Technical University, 1 Gagarin Square, Rostov-on-Don 344000, Russian Federation
yevgeniya.medvedkina@gmail.com

Abstract. The research paper discusses modern digital technologies in the current conditions of global economy in the financialization. The existing theoretical and methodological basis argues that modern digital technologies lead to stimulating the economic growth of the country. This study attempts to explore this statement using an econometric model, which allows us to get an idea of future development opportunities in the case of the use of innovative tools and technologies that are successful in Russia and Japan. The article analyses indicators such as GDP growth (%), mobile-phone users (per 100 adults), Bitcoin trading volume (operation with JPN/RUB to Bitcoin), numbers of traders per minute annually, share of non-cash operation (%), e-money payment transactions (millions), ATMs (per 100,000 adults). The research paper presents regression-correlation analysis with using of these indicators. The article confirms the lack of influence of fintech on Japan and on Russia.

Keywords: Financialization · Digital and Financial Technologies · Financial System · Economic Growth · Fintech · Financial Fervices

1 Introduction

The rapid progress in the field of financial technologies (fintech) leads to the transformation of the financial and economic landscape, which opens up a wide range of opportunities and at the same time creates potential risks. Financial technologies can contribute to potential economic growth and poverty reduction by enhancing financial development as well as improving the availability and efficiency of financial services, but they can also create risks for consumers and investors and, more generally, stability and integrity of the financial system.

National governments are interested in enhancing the positive effects and mitigating the potential risks of fintech. Currently, many international and regional organizations are exploring various aspects of financial technology under their powers.

The interdisciplinarity, the ambiguity of the predicted results obtained, as well as the significant dependence of various empirical models on exogenous factors, form the stochasticity of the estimates of not only the obtained results but also the learning process itself with its disposable tools.

A. Gibadullin (Ed.): DITEM 2023, LNNS 942, pp. 118–130, 2024.
https://doi.org/10.1007/978-3-031-55349-3_10

Modern digital and financial technologies in the financialization of the global economy, the author mostly managed to solve one of the fundamental scientific problems of interdisciplinarity to combine the methodology of general philosophical, social and human sciences with the methods of economic and mathematical modeling, which was reflected in using the methods of correlation and regression analysis to identify the dependence of the distribution fintech, as well as the degree of its influence on the economic growth of the state.

The depth and breadth of the retrospective analysis of the special scientific literature within the scope of the subject field of the research make it possible to argue that the results of the research, namely, the study of the emergence and development of financial technologies, blockchains with the development of a qualitative new empirical and methodological approach to determine and predict the bilateral impact on economic growth based on the analysis of qualitative and quantitative indicators indicate the reliability of the results of scientific research.

The aim of the article is to study modern digital and financial technologies in the current conditions of global economy and to create an econometric model in order to make a perspective of the future development opportunities in case of application of innovative tools and technologies successful in Russia and Japan.

2 Materials and methods

This study is based on the works of Schumpeter J. [1] and Kondratiev N. [2], who were investigated the essence and role of innovations is first determined.

It should be noted that scientists understand capital as materials, machinery, equipment and money, land is a physical space, as well as natural resources, and labor is workers. Information is a rather broad concept that includes the media, such as TV, radio and also data. Technology means computers, machines and robots, but also the entire body of knowledge or science that informs or improves a production process.

The first industrial revolution also known as the "great industrial revolution" became the transition from manual to machine labor. The relocation of impoverished peasants to the cities, the ruin of small artisans provided the rapidly growing industry with labor and contributed to the expansion and acceleration of urbanization [3]. Land and technology faded into the background. Such a factor as information was just beginning to emerge. These data are supported by empirical calculations, which are presented in the works of Sturgill and Giedeman [4].

The second industrial revolution began in the second half of the 19th century and spread to Western Europe, the USA, the Russian Empire and Japan. It lasted until the beginning of the twentieth century and was characterized by the development of in-line production, the extensive usage of electricity and chemicals. Mass in-line production began with the conveyor production of such an innovative product for that time like a car. Then, it spread to other industrial sectors. It changed not only the production itself, but also its management system.

Also during this period, the first power stations were built, telephone and radio were created. The acceleration of economic growth and the rapid growth of labor productivity, and consequently of its wages, ensured a qualitative leap in the standard of living of the population, but at the same time unemployment also increased [3].

From this it follows that capital and labor continued to be leaders, but labor began to decrease in importance. Moreover, land became less necessary for economic growth. But technology and information began to develop actively.

The third industrial revolution is the transition in production to the use of information and communication technologies, the priority use of renewable energy sources and the formation of a post-industrial society. The growth of labor productivity in all types of labor has led to a decrease in demand for both workers in routine positions and engineering and technical workers and clerks. In business management, centralization has weakened, but horizontal ties have intensified.

All these changes formed the prerequisites for the next revolution, which not only was characterized by the emergence of new technologies, but also started to modify in the contours of the financial system [5].

It should be noticed that all these processes occurred with varying intensity in developed countries, countries with catching-up economies and developing countries. At the same time, the "leading countries" have already begun the transition to the so-called Industry 4.0.

The fourth industrial revolution began in the 2010s and continues today. The most significant for the fourth industrial revolution were such technological breakthroughs as artificial intelligence, robotization, nano – and biotechnology, new ways of accumulating and storing information. The processes of production, exchange and consumption occur through the collection, processing and transmission of large amounts of information. From the point of view of the global financial and economic system in the fourth industrial revolution, the most significant was the emergence in 2009 of such innovative technologies as a blockchain in which is applicable both in the financial industry and in other industries [2].

During the first part of the fourth industrial revolution the capital was the leader, the next stage was occupied by labor, information and technology increasingly growing and land was in the last position (see Fig. 1).

It can be noted that from one revolution to another, not only the priority in the factors of production has changed, but also there are changes in their forms. Thus, cheap unskilled labor was replaced by more expensive and skilled. Capital which was originally more connected with factories and equipment, now implies the notion of finance.

We can also correlate industrial revolution with economic Kondratiev's and Schumpeter's cycles. These scientists made a huge contribution to economics with their ideas about the impact of innovation on economic growth. For example, the last 6th Kondratiev cycle is also associated with AI, robots, alternative energy sources, nanotechnologies, which confirms our theory that the contribution of capital, technology, information and labor as factors of production will grow.

This study also will perform a regression analysis using the STATISTICA statistical package and will provide an economic interpretation of the results. To identify general trends in achieving sustainable development goals on a planetary scale, a sample of 6 annual indicators of GDP growth (%), mobile-phone users (per 100 adults), Bitcoin trading volume (operation with JPN/RUB to Bitcoin), numbers of traders per minute annually, share of non-cash operation (%), e-money payment transactions (millions), ATMs (per 100,000 adults). The presented sample of key indicators is a cross-country

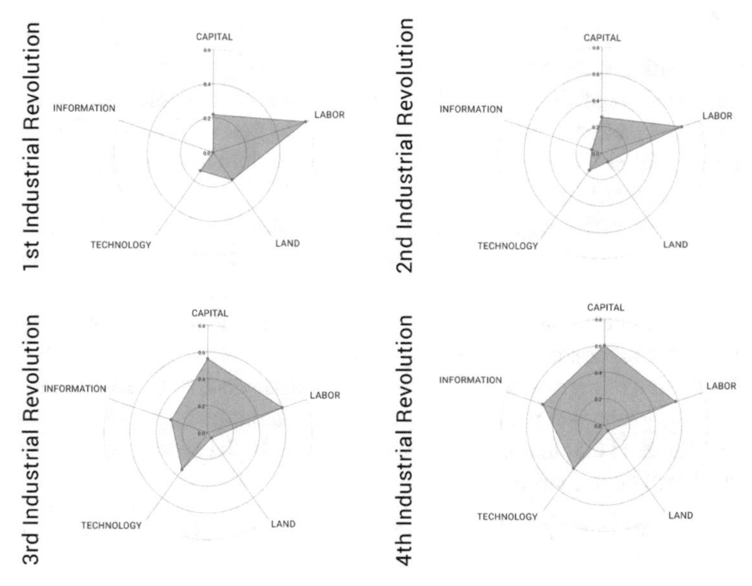

Fig. 1. Factors of production according to industrial revolutions.

comparative analytics for the Russian Federation and Japan for the period from 2011 to 2017.

Analytical research tools assume the implementation of the following successive operations using the statistical analysis package STATISTICA:

Stage 1. The choice of factors and response regression analysis. It is carried out on the basis of ideas about the nature of the problem being studied, the intuition of a specialist or the experience of similar studies.
Stage 2. Search for multicollinear factors of regression analysis. The factors are called multicollinear if there is a fairly strong correlation between them. This problem makes it difficult to rank factors by the degree of influence on the response. It is recommended to remove the multicollinear factors from the model if this is not essential for solving the set task.
Step 3. Study of the relative importance of multicollinear regression factors. The relative strength of the influence of factors on the response is shown by standardized regression coefficients (Beta). Of the two multicollinear factors, one that has a lower Beta is excluded from the analysis.
Step 4. Analysis of residual regression analysis. Residuals are the differences between the actual response values and the values predicted by the regression equation for the same factors.
Step 5. Analysis of the regression equation and removal of factors that do not affect the response of the regression analysis. Factors with p greater than 0.05 can be excluded from the analysis, i.e. they do not significantly affect the response. After any exception, the entire preceding analysis algorithm must be repeated. If there are several factors with a significance level of more than 0.05, the one with the significance level above the

others is first removed. The analysis is repeated first and only after that the next factor is considered with p greater than 0.05.
Stage 6. Assessment of the acceptability of the model as a whole regression analysis According to the analysis of variance analysis table (ANOVA) $p = 0.000$ ... less than 0.05, then the prediction error for the constructed model will be less than with the "naive" forecast, i.e. model can be considered acceptable.
Step 7. R^2 analysis of the regression analysis R^2 – coefficient of determination, shows the proportion of response variability that occurs under the simultaneous influence of all factors included in the model. The more R^2, the higher the quality of the model. A small value of R^2 may indicate inadequate selection of factors and indicates the inexpediency of making forecasts for such a model. adjusted R^2.
Stage 8. Prediction of regression analysis. To build a forecast, it is necessary to enter the predicted values of the factors, the influence of which on the response is established. It must be remembered that the more accurate the forecast, the closer the predicted values of the factors will be to their average.

The data sources are such statistical resources as World Bank [6] and Bitcoin [7].

3 Research Results

The industrial revolution could not happen without the presence of the necessary factors of production. Traditionally, there are three main factors of production, such as capital, land and labor. However, in order to consider which of the factors were keys to each revolution, we will expand the traditional model and add two modern factors. Paul Romer [8] received the Nobel Prize for his work, in which he proved the importance of another indicator of economic growth, which is technology in 2018. Also, since we live in an information society, we cannot deny the importance of such a factor as information. Thus, we explore each stage of the development of the world economy using five factors of production: traditional such as capital, land, labor and modern such as technology and information.

As part of the research goal setting, a sample of 6 variables (var_1, ... var_6) was formed for Japan and the Russian Federation for the period from 2011 to 2017 (Table 1, Table 4), which will be used as a matrix for analysis.

After formalizing the set of factors and the logical definition of the response (Stage 1), it is advisable to search and analyze multicollinear factors of regression analysis. To this end, descriptive statistics were calculated in the STATISTICA package and the most significant factors were identified, the results of which are given in Table 2.

A strong correlation is observed in 5 of 6 independent variables of the main characteristics of fintech, therefore multicollinearity can be observed and for its identification and pre-emption of influence it is necessary to construct standardized Beta-coefficients.

Based on the above calculations of the Beta – coefficients, two negative values are observed for % 100 of internet users (var_1) and Share of non-cash operation (var_5). To eliminate multicollinearity from the set of independent variables of the main characteristics of fintech in Japan, it is necessary to remove the one for which according to Beta is the smallest coefficient, that is, Share of non-cash operation (var_5).

Table 1. Dynamics of distribution of the main characteristics of fintech in Russia [6, 7].

Year	GDP growth, %	% 100 of internet users	% 100 mobile-phone users	Bitcoin trading volume [BTC], Operation with RUS to Bitcoin)	Numbers of traders per minute × 60 × 24 × number of days per each month (Traders – RUS to Bitcoin)	Share of non-cash operation (%)	E-money payment transactions (millions)	ATMs (per 100,000 adults)
	Y	var_1	var_2	var_3	var_4	var_5	var_6	var_7
2011	5.28	49	142.22	26448.94	974	19	106.19	116.89186
2012	3.65	63.8	145.07	729.4256	463	23	225.67	142.00364
2013	1.78	67.97	152.02	12728.67	3505	27	564.38	156.50616
2014	0.73	70.52	153.75	10174.64	19250	33	1013.58	185.32424
2015	−2.82	70.1	157.96	187974.6	902307	40	1039.99	172.60456
2016	−0.22	73.09	159.15	344330.3	2033153	47	1279.38	168.69774
2017	1.64	76.01	157.86	277250.2	4333686	58	1358.37	163.92613

Table 2. Correlation matrix of independent variables of the main characteristics of fintech in Russia.

Variable	Correlations						
	% per 100 mobile-phone users	Bitcoin trading volume [BTC] (Operation with RUB to Bitcoin)	Numbers of traders per minute x 60 x 24 x number of days per each month (Traders RUB-Bitcoin)	Share of non-cash operation (%)	E-money payment transactions (millions)	ATMs (per 100,000 adults)	GDP growth (%)
% 100 mobile-phone users	1	0.771865	0.628658	0.87244	0.962198	0.842541	-0.862098
Bitcoin trading volume [BTC] (Operation with RUB to Bitcoin)	0.771865	1	0.824236	0.878202	0.794426	0.368505	-0.532047
Numbers of traders per minute x 60 x 24 x number of days per each month (Traders RUB-Bitcoin)	0.628658	0.824236	1	0.921996	0.734411	0.275271	-0.244939
Share of non-cash operation (%)	0.87244	0.878202	0.921996	1	0.935153	0.606412	-0.574173
E-money payment transactions (millions)	0.962198	0.794426	0.734411	0.935153	1	0.821384	-0.743679
ATMs (per 100,000 adults)	0.842541	0.368505	0.275271	0.606412	0.821384	1	-0.82165
GDP growth (%)	-0.862098	-0.532047	-0.244939	-0.574173	-0.743679	-0.82165	1

The next step is to recalculate descriptive statistics and identify the most significant factors. The next step is to recalculate the standardized Beta – coefficients, the results of which are shown in Table 3.

Table 3. Correlation matrix of independent variables of the main characteristics of fintech in Russia, excluding share of non-cash operations (var_5).

Variable	Correlations					
	% per 100 mobile-phone users	Bitcoin trading volume [BTC] (Operation with RUB to Bitcoin)	Numbers of traders per minute x 60 × 24 × number of days per each month (Traders RUB-Bitcoin)	E-money payment transactions (millions)	ATMs (per 100,000 adults)	GDP growth (%)
% 100 mobile-phone users	1	0.771865	0.628658	0.962198	0.842541	−0.862098
Bitcoin trading volume [BTC] (Operation with RUB to Bitcoin)	0.771865	1	0.824236	0.794426	0.368505	−0.532047
Numbers of traders per minute × 60 × 24 × number of days per each month (Traders RUB-Bitcoin)	0.628658	0.824236	1	0.734411	0.275271	−0.244939
E-money payment transactions (millions)	0.962198	0.794426	0.734411	1	0.821384	−0.743679
ATMs (per 100,000 adults)	0.842541	0.368505	0.275271	0.821384	1	−0.82165
GDP growth (%)	−0.862098	−0.532047	−0.244939	−0.743679	−0.82165	1

Based on the above calculations of the Beta – coefficients, two negative values are observed for Bitcoin trading volume (var_3) and ATMs per 100,000 adults (var_7). However, significance indicators (p-value) are greater than 0.05, which allows us to accept the presented final set of independent variables (var_3, var_4, var_6, var_7) as the most significant and explaining 83.4% of the influence on the response.

The presented data analysis allows returning to the verification of the hypothesis, which is financial technologies and the process of financialization have a strong interdependence with economic growth and stable economy in the conditions of economic and mathematical modeling.

The next step is the calculation of residuals, which are the differences between the actual values of the response and the values predicted by the regression equation for the same factors (see Fig. 2).

As can be seen from the graph, the distribution of residuals is normal except for the interval of the confidence interval [−0.5; 0], which can be explained by an insignificant observation period. This does not reduce the significance of the model.

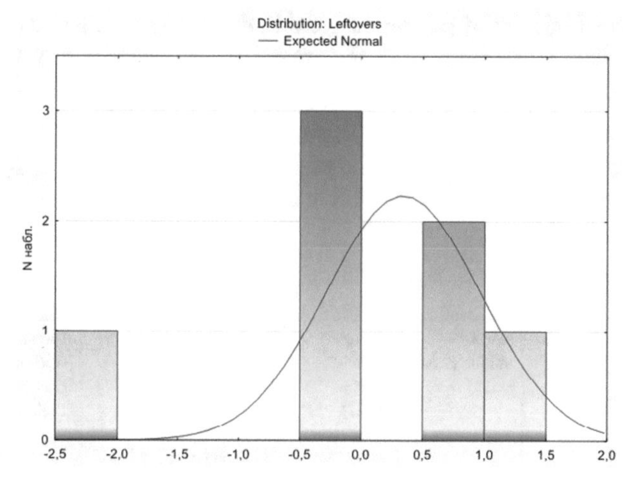

Fig. 2. Graph of the distribution of residues of independent variables of the main characteristics of fintech in Russia.

As a result, the regression equation of independent variables of the main characteristics of fintech in Russia for var_3, var_4, var_6, var_7 will have the form (formula 1):

$$\begin{aligned} \text{GDP}_{growth}(\%) &= 19.921 - 0.00002 \times \text{Bitcoin trading volume } [BTC] \\ &+ 0.0000 \times \text{Numbers of traders per minute} \times 60 \times 24 \times \text{numbers of days per each month} \\ &+ 0.0298 \times \text{E-money payment transactions\_}(millions) \\ &- 0.1233 \times \text{ATMs } (per\ 100\ 000\ adults) \end{aligned} \quad (1)$$

The presented economic and mathematical analysis of the influence of independent variables that characterize the distribution of fintech in Japan, namely: Mobile-phone users, Bitcoin trading volume, Numbers of traders per minute annually, share of non-cash operation, e-money payment transactions (millions), ATMs per 100,000 adults to response of GDP growth (%) allows you to do the following findings:

- The importance of fintech to achieve economic growth in Russia is insignificant, since only one of the stated indicators (E-money payment transactions (millions)) affects the amount a little more than the statistical error of the resulting economic and mathematical model;
- It should be noted that such indicators as Bitcoin trading volume [BTC], Operation with RUB to Bitcoin) and ATMs (per 100,000 adults) show an inverse relationship, and therefore it can be hypothesized that Russia's economic growth is largely associated with the monetary aggregate M_0 (cash), affecting the reduction in the money supply in circulation, targeting inflation at the level of 2.5–2.75% and, as a result, concentration of money in the banking sector, which is an inhibitor of economic development;
- Independent variables such as Numbers of traders per minute annually (Traders RUB-Bitcoin) and E-money payment transactions (millions) have a direct relationship to GDP growth (%), but their significance indicates about the peculiarities of state policy

and the lack of interest of private business in the distribution and use of fintech in economic relations and some conservativeness of the financial and economic policy of the state.

As part of the research goal setting, a sample of 6 variables (var_1, ... Var_6) was formed for Japan and the Russian Federation for the period from 2011 to 2017 (Table 4), which will be used as input data for building a regression model.

Table 4. Dynamics of the distribution of the main characteristics of fintech in Japan [6, 7].

Year	GDP growth, %	% 100 of internet users	% 100 mobile-phone users	Bitcoin trading volume [BTC], Operation with JPN to Bitcoin)	Numbers of traders per minute × 60 × 24 × number of days per each month (Traders JPN-Bitcoin)	Share of non-cash operation (%)	E-money payment transactions (millions)	ATMs (per 100,000 adults)
	Y	var_1	var_2	var_3	var_4	var_5	var_6	var_7
2011	−0.11542	79.05	103.31	1542.17	94	14.1	2342.05	128.5757
2012	1.49509	79.5	109.89	106154	10580	15.1	2836.59	127.896
2013	2.00026	88.22	115.26	429459	105936	15.3	3453.42	128.2963
2014	0.37471	89.11	123.17	300829	118323	16.9	4039.55	127.4941
2015	1.35382	91.06	125.46	428623	799890	18.2	4678.45	127.6515
2016	0.93819	93.18	130.61	5704661	5384201	20	5191.6	127.7969
2017	1.73455	90.87	135.54	6658475	17938464	21.3	6320	127.767

After formalizing the set of factors and the logical definition of the response (Stage 1), it is advisable to search and analyze multicollinear factors of regression analysis. To this end, descriptive statistics were calculated in the STATISTICA package and the most significant factors were identified, the results of which are given in Table 5.

A strong correlation is observed in 3 of 6 independent variables of the main characteristics of fintech, therefore multicollinearity can be observed, and to identify and anticipate the effect, it is necessary to carry out n − 1 iteration to calculate standardized Beta – factors that best describe the regression.

Based on the above calculations of the Beta – coefficients, one negative values are observed for Bitcoin trading volume [BTC], Operation with JPN to Bitcoin) (var_3). However, significance indicators (p-value) are greater than 0.05, which allows us to accept the presented final set of independent variables (var_3, var_4, var_6, var_7).

It should be noted that the presented set of independent variables in terms of significance meets the tabular verification requirements, but they explain only 22.1% of the influence on the response, that is, GDP growth (%). This circumstance casts doubt on the validity of the influence of independent variables on the response.

The next step is the calculation of residuals, which are the differences between the actual values of the response and the values predicted by the regression equation for the same factors (see Fig. 3).

Table 5. Correlation matrix of independent variables of the main characteristics of fintech in Japan.

Variable	Correlations							
	% per 100 of Internet users	% per 100 mobile-phone users	Bitcoin trading volume [BTC] (Operation with RUB to Bitcoin)	Numbers of traders per minute x 60 x 24 x number of days per each month (Traders RUB-Bitcoin)	Share of non-cash operation (%)	E-money payment transactions (millions)	ATMs (per 100,000 adults)	GDP growth (%)
% per 100 of Internet users	1	0.905947	0.59593	0.436103	0.811885	0.847333	-0.611404	0.336458
% per 100 mobile-phone users	0.904557	1	0.785067	0.71367	0.96763	0.981739	-0.72471	0.366141
Bitcoin trading volume [BTC] (Operation with RUB to Bitcoin)	0.59593	0.785067	1	0.886177	0.886422	0.846287	-0.283302	0.260967
Numbers of traders per minute x 60 x 24 x number of days per each month (Traders RUB-Bitcoin)	0.436103	0.71367	0.886177	1	0.813232	0.819707	-0.247929	0.343766
Share of non-cash operation (%)	0.811855	0.96763	0.886422	0.813232	1	0.98632	-0.62308	0.313651
E-money payment transactions (millions)	0.847333	0.981739	0.846287	0.819707	0.98632	1	-0.6234	0.391434
ATMs (per 100,000 adults)	0.611404	0.72471	-0.283302	-0.247929	-0.62308	-0.6234	1	-0.176053
GDP growth (%)	0.336458	0.366141	0.260967	0.343766	0.313651	0.391434	-0.176053	1

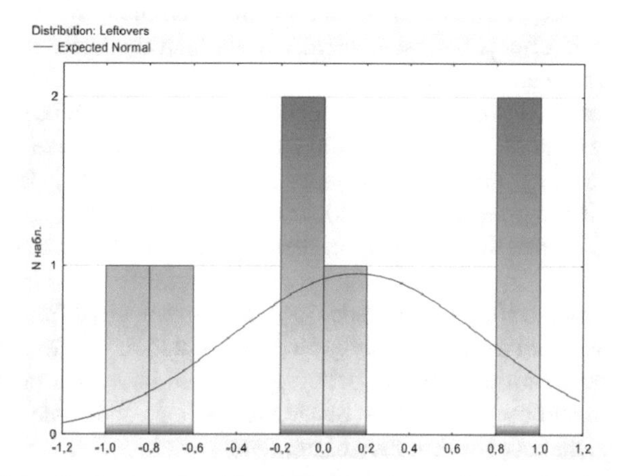

Fig. 3. Graph of the distribution of residues of independent variables of the main characteristics of fintech in Japan.

As can be seen from the graph, the distribution of residuals is not normal and does not correspond to the theoretically possible, which confirms the earlier suggestion that there is no dependence of the response on the independent variables included in the model.

As a result, the equation of potential regression of independent variables of the main characteristics of fintech in Japan for var_3, var_4, var_6, var_7 will look like (formula

2):

$$
\begin{aligned}
\mathrm{GDP}_{growth}(\%) = & -68.63621 - 0.0000 \times \text{Bitcoin trading volume } [BTC] \\
& + 0.0000 \times \text{Numbers of traders per minute} \times 60 \times 24 \times \text{numbers of days per each month} \\
& + 0.0005 \times \text{E-money payment transactions\_}(millions) \\
& - 0.5311 \times \text{ATMs } (per\ 100\,000\ adults)
\end{aligned}
\tag{2}
$$

4 Discussion of Results

The presented economic and mathematical analysis of the influence of independent variables that characterize the distribution of fintech in Japan, namely: Mobile-phone users, Bitcoin trading volume [BTC] (Operation with JPN to Bitcoin), Numbers of traders per minute annually (Traders JPN-Bitcoin), Share of non-cash operation (%), E-money payment transactions (millions), ATMs (per 100,000 adults) to response – GDP growth (%) allows you to do the following findings:

- It should be recognized that among the independent significance of fintech to achieve economic growth in Japan is extremely insignificant, since only one of the stated indicators (ATMs (per 100,000 adults)), which is indirectly a characteristic of the spread of fintech, affects the value slightly more than the statistical error obtained economic and mathematical model;
- It should be noted that the indicator Bitcoin trading volume [BTC], Operation with JPN to Bitcoin) and ATMs (per 100,000 adults) shows an inverse relationship, and therefore it can be hypothesized that Japan's economic growth is largely associated with the monetary aggregate M0 (cash), affecting the devaluation of the yen and the growth of public debt, which in 2018 exceeded 250% of GDP and ranked first in the world;
- Independent variable Numbers of traders per minute annually (Traders JPN-Bitcoin) and have a direct relationship to GDP growth (%), but their insignificant significance indicates the peculiarities of public policy and lack of interest of private business in the distribution and use of fintech in economic relations and some conservativeness of the financial and economic policy of the state.

Based on the analysis of the effect of independent variables on the response for Japan and Russia, the final stage of comparative studies is a comparison of the regression coefficients of independent variables of fintech in Russia and Japan, which are presented in Table 6.

Table 6. Comparative analysis of the regression coefficients of independent variables distribution fintech in Russia ($R^2 = 0{,}83476$) and Japan ($R^2 = 0{,}220702$).

Country	Bitcoin trading volume [BTC], Operation with RUB/JPN to Bitcoin)		Numbers of traders per minute × 60 × 24 × number of days per each month (Traders RUB/JPN-Bitcoin)		E-money payment transactions (millions)		ATMs (per 100,000 adults)	
	Dependence	value	dependence	value	dependence	value	Dependence	Value
Russia	Reverse	0.00002	direct	0.0000	direct	0.00298	Reverse	0.1233
Japan	Reverse	0.0000	direct	0.0000	direct	0.0005	Direct	0.5311

5 Conclusion

Based on the analysis performed, the following final conclusions can be drawn:

- For both Japan and Russia, the significance of the factors characterizing the penetration, distribution and participation of Fintech in economic relations is extremely insignificant and has a weak effect on GDP growth over the period from 2011 to 2017;
- It should be noted that despite the weak representativeness of the independent variables in the response, the presented sample of the most common worldwide factors best describes the situation with fintech in Russia, and for Japan it is necessary to conduct additional studies that will reveal additional national features and forms of fintech that provide significant impact on GDP growth;
- From another point of view, the obtained results may indicate some protectionism of the financial system of Japan in relation to modern fintech, which speaks of conservative instruments of state economic policy to achieve non-inflationary economic growth;
- It is advisable to note the almost complete equality appear standardized Beta – coefficients.

Regression-correlation analysis confirms the lack of influence of fintech on the economy of developed countries (Japan) and on the economy of the transitive type (Russia). Although for the final confirmation or refutation of these theses, additional research is needed.

References

1. Schumpeter, J.: The Theory of Economic Development: An Inquiry into Profits, Capital, Credit, Interest, and the Business Cycleed. Harvard University Press, Cambridge, MA (2008)
2. Kondratiev, N.: The long waves in economic life. Rev. Econ. Stat. **17**, 101–115 (2017)
3. Urinson, J.M.: Industrial Revolution and Economic Growth. Liberal mission, Moscow (2018)
4. Sturgill, B., Giedeman, D.: Factor Shares, Economic Growth, and the Industrial Revolution, EBHS, pp. 23–25 (2016)
5. World Economic Forum, Homepage, https://www.weforum.org/agenda/2016/01/the-fourth-industrial-revolution-what-it-means-and-how-to-respond, Last accessed 5 Nov 2023

6. Data World Bank Homepage, https://databank.worldbank.org. Last accessed 5 Nov 2023
7. Data Bitcoin trading volume. Homepage, https://data.bitcoinity.org/markets/volume/30d?c=e&t=b. Last accessed 5 Nov 2023
8. Romer, P.: Endogenous technological change. J. Polit. Econ. **98**(5), 71–102 (1990)

Simulation of a Tubular Pyrolysis Reactor Using Comsol Multiphysics Software

Gulom Uzakov[1], Sayyora Mamatkulova[1](✉), Farrux Qodirov[2], and Ochilova Sojida[3]

[1] Karshi Engineering Economic Institute, 225, Mustaqillik Street, Karshi, Uzbekistan 180100
urisheva80@mail.ru

[2] University of Economics and Pedagogy Non-State Educational, 13, I.A. Karimov Street, Karshi, Uzbekistan 180100

[3] Karshi Branch of Tashkent University of Information Technologies Named After Muhammad Al-Khwarizmi, Beshkent Road 3 Km, Karshi, Uzbekistan 180100

Abstract. This article provides a visual realization of the scientific study of the process of decomposition of biomass at high temperatures in the absence of oxygen in the tubular reactor of pyrolysis plant (PP) which is modeled with the help of Comsol Multiphysics software. The developed model of tubular reactor PP in simplified geometry allows to significantly reduce the amount of computing work, calculation time by 72% and is a tool for research and optimization of pyrolysis process. The model is a three-dimensional representation of the degradation process of plant biomass which has a specific thermal conductivity, density and dynamic viscosity. The study emphasizes the importance of using computer modelling in the study and optimization of pyrolysis processes. This opens up new opportunities for improving pyrolysis and deriving valuable biomass products. Moreover, the results of the study could be used to improve existing pyrolysis plants and to develop new technologies for biomass processing. Further studies are planned to further refine the model and expand its application.

Keywords: tubular reactor · Comsol Multiphysics · 3D modeling · biomass · pyrolysis

1 Introduction

A feature of modern scientific methods for studying complex physicochemical processes and technological systems is the creation of models for describing processes and predicting changes in the state of the systems under study. The tubular reactor of a pyrolysis plant (PP) is considered as an object of modeling, which is the most important component of the production of biofuels based on pyrolysis, designed to implement the necessary thermal chemical processes [1, 2].

A tubular reactor is one of the most common types of reactors used in PP. Pyrolysis is a chemical process in which organic materials decompose at high temperatures in the absence of oxygen. This process is used to produce valuable products such as biogas, pyrolysis liquid and solid fuel [2, 3].

A. Gibadullin (Ed.): DITEM 2023, LNNS 942, pp. 131–142, 2024.
https://doi.org/10.1007/978-3-031-55349-3_11

However, effective and safe management of these reactors requires an accurate understanding of the processes occurring inside them. This is where computer modeling with the help of application software comes into play. Computer simulation of a PP tubular reactor is an important task in the field of renewable energy and thermal processes.

With the help of computer modeling, it is possible to create a detailed three-dimensional model of a tubular reactor of a pyrolysis plant and conduct a series of virtual experiments to see how the reactor will react to various conditions. This can help us optimize the design of the reactor, improve its efficiency and safety, and save time and money on conducting physical experiments.

The importance of this problem is emphasized by the fact that the results of these simulations can be used to improve pyrolysis processes at the industrial level, which in turn can lead to more efficient energy production and reduce environmental impact. In this context, the use of the Comsol Multiphysics software package for modeling a PP tubular reactor represents a significant contribution to this field.

Modeling and optimization of chemical processes are based on well-developed mathematical methods included in Comsol Multiphysics software packages.

2 Methodology

Comsol Multiphysics is a powerful modeling tool that allows you to formulate, analyze and edit chemical reaction equations, as well as equations, functions and variables describing reaction kinetics. It also offers users advanced mathematical and numerical methods adapted for the calculation of chemical systems [4].

With the help of Comsol Multiphysics software, engineers and scientists model structures, devices and processes in all areas of engineering, manufacturing and scientific research. Using the Comsol Multiphysics platform, it is possible to analyze both individual and interrelated physical processes. The Model Builder environment, which allows you to go through all the stages from building a geometric model, setting the properties of materials and describing the physics of the problem to performing calculations and analyzing the simulation results.

Having developed a model, it is possible to create a simulation application based on it in the Application Builder Environment with a specialized interface for solving typical tasks for a wide range of users, including colleagues, clients and people with minimal experience in numerical modeling. For efficient and structured storage of models and applications, the Comsol Multiphysics platform contains a Model Manager System, which is an environment for efficiently storing models in a database, monitoring and managing various versions of models and associated files.

To create models for solving specialized applied and engineering problems, you can supplement the capabilities of Comsol Multiphysics software with expansion modules in any combination. They provide access to additional specialized tools, but at the same time they are available in a single user interface and function together as a whole, regardless of what physical phenomena are being modeled [5].

The purpose of modeling a tubular PP reactor at Comsol Multiphysics is to study and optimize the processes occurring inside the reactor. Using the Comsol Multiphysics software, you can perform the following tasks:

- Analysis of thermal processes: Study of the temperature distribution inside the PP tubular reactor, the influence of thermal processes on the speed and efficiency of pyrolysis.
- Chemical reaction research: Modeling of chemical reactions occurring in a PP tube reactor and their effect on the final product.
- Optimization of PP tubular Reactor design: Using simulation results to optimize the design of a tubular reactor to improve its efficiency and safety.
- Flow modeling: Study of material flows inside a tubular reactor and their effect on the pyrolysis process.

It is important to note that the effective use of Comsol Multiphysics will require a deep understanding of the physical and chemical processes occurring inside the PP tubular reactor, as well as experience with numerical methods and modeling software.

The pyrolysis process in a tubular reactor is carried out by heat, so the heat transfer conditions determine the entire process. The geometry of the system (tubular reactor), in particular the geometry and size of the reacting biomass particles, strongly affect the heat exchange. In the pyrolysis process, many different reactions take place, involving substances originally present in the raw materials, as well as intermediates and end products [6–9].

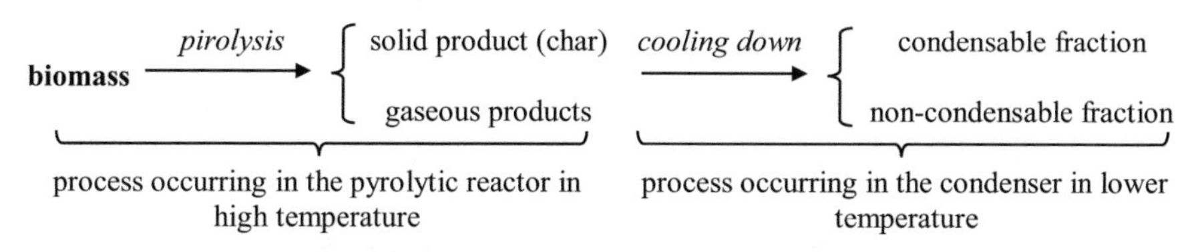

Fig. 1. Diagram of the pyrolysis process [6].

A tubular pyrolysis reactor is a device used for the thermal treatment of carbon-containing waste. The pyrolysis process takes place inside the reactor, which is the thermal decomposition of organic and many inorganic compounds (Fig. 1).

The PP tubular reactor we are modeling is made of stainless heat-resistant steel (Fig. 2.). It consists of several blocks: a pyrolysis reactor, a water jacket, a burner device, a pipeline of a vapor-gas mixture. All units are mounted on a single metal frame. It is important to note that the design and operation of a tubular reactor may vary depending on the specific pyrolysis plant and process requirements.

The following physical processes take place in the tubular reactor of the biomass pyrolysis plant [10–12]:

- Thermal decomposition: Pyrolysis of biomass is the thermal decomposition of organic compounds. During pyrolysis, chemical reactions occur, as a result of which hydrocarbons decompose into lighter molecules or chemical elements.
- Radiation heating: In the radiation section there are tubular pyrolysis reactors heated by the heat of combustion of externally supplied combustible gas in the burners of this section.
- Heat exchange: A heat exchange process also takes place in a tubular reactor, which affects the temperature field inside the reactor.

- Chemical reactions: The mathematical model of the pyrolysis process is based on the use of a multi-stage decomposition scheme of chemical components of biomass, implemented taking into account the residence time of the resin in the layer of hot coke and the temperature of the layer.

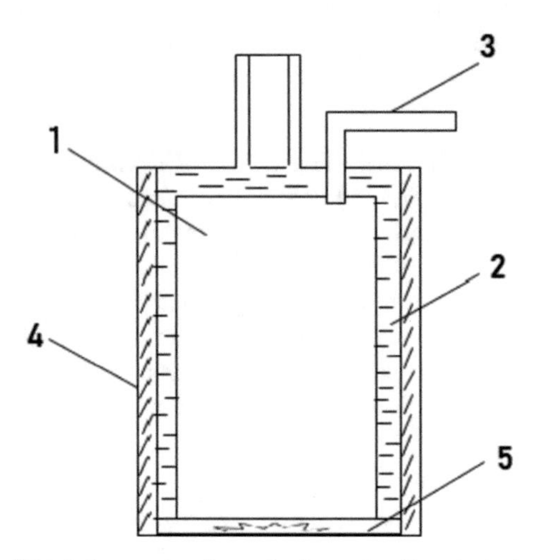

Fig. 2. Diagram of the PP tubular reactor: 1-pyrolysis reactor; 2-water jacket; 3-steam-gas mixture pipeline; 4-insulation; 5-burner device.

In this paper, the process of heat exchange in a tubular PP reactor is stimulated. A flowchart has been developed for modeling a PP tubular reactor and visualizing thermal processes, which contains the stages of the process.

The developed block diagram (Fig. 3) describes the process of modeling a tubular reactor of a pyrolysis plant using Comsol Multiphysics software. The circuit consists of eight blocks arranged vertically and connected by lines, which indicates the sequence of the process:

- Physics Setting: This stage involves determining the physical parameters to be used in the model, such as thermal conductivity, density and dynamic viscosity.
- Parameter definition: Here specific parameters of the model are defined, such as initial and boundary conditions.
- Geometry creation: At this stage, the three-dimensional geometry of the tubular reactor is created.
- Definition of Equations (PDE): Here are formulated equations describing the physical processes inside the reactor, such as heat and mass transfer.
- Final model: After determining all the parameters and equations, a final model is obtained that can be used for numerical experiments.
- Grid Setup: At this stage, a grid is determined for the numerical solution of the model equations.
- Equation Solving: Here the model equations are solved using the selected numerical method.

- Analysis of results: Finally, the simulation results are analyzed and interpreted to obtain useful information about the pyrolysis process in a tubular reactor.

This flowchart is a useful tool for modeling a PP tubular reactor, as well as in the study and optimization of the pyrolysis process, allowing researchers to predict the behavior of the system under various conditions and determine the optimal parameters for maximum efficiency.

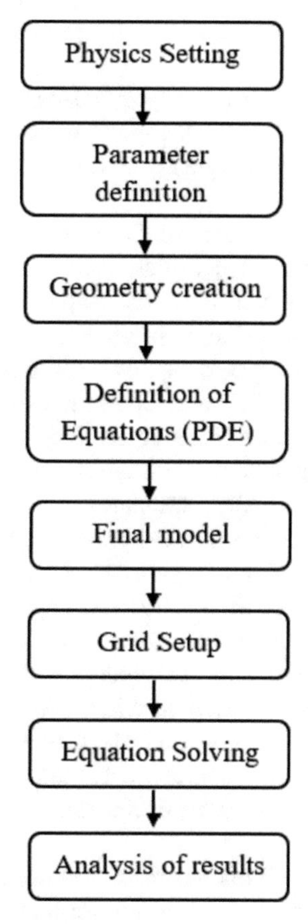

Fig. 3. Block diagram of Comsol Multiphysics software used to model a PP tubular reactor.

With the help of Comsol Multiphysics software, it is possible to analyze both individual and interrelated physical processes. It includes a model development environment that allows you to go through all the stages from building a geometric model, specifying the properties of materials and describing the physics of the problem to performing calculations and analyzing the simulation results.

After developing the model, you can create a simulation application based on it in the Application Development Environment with a specialized interface for solving typical tasks for a wide range of users, including colleagues, clients and people with minimal experience in numerical modeling.

The advantage of modeling using Comsol Multiphysics software is the simplicity of the user interface, which helps to reduce the amount of complex computer coding. Another advantage of Comsol Multiphysics software is its flexibility in modeling complex geometries and combining various physical processes. Comsol Multiphysics software also has many built-in tools for data postprocessing that make it easy to study various quantities [13].

All modeling and simulation processes specified in the flowchart (Fig. 3) were carried out using Comsol Multiphysics 6.1 on a computer with an 11th Gen Intel(R) Core(TM) i5-1135G7 processor with a clock frequency of 2.40 GHz, 16 GB of RAM and a 64-bit operating system.

To start modeling a tubular reactor, we chose the following types of physics: Non isothermal Pipe Flow, Heat Transfer in Solids.

Non-isothermal Pipe Flow – Non-isothermal pipe flow interface is used to calculate temperature, velocity and pressure fields in pipes and channels of various shapes. It approximates the flow profile in a pipe using 1D assumptions in curve segments or lines. These lines can be drawn in 2D or 3D and represent a simplification of hollow pipes.

Heat Transfer in Solids – heat transfer at the interface of solids is used to simulate heat transfer due to thermal conductivity, convection and radiation. The solid-state model is active by default in all domains. All functionality is also available to enable other types of domains, such as the fluid domain. The temperature equation defined in solid regions corresponds to the differential form of Fourier's law, which may contain additional contributions, such as heat sources. Stationary modeling, modeling in time and frequency domains is supported in all spatial dimensions.

To simulate the process, the Time Dependent type of training is selected. Time Dependent is a time–dependent study used when field variables change over time. In heat transfer, it is used to calculate temperature changes over time.

With the help of Geometry 1 resources, a PP tubular reactor was modeled in 3D format (Fig. 4), which consists of a pyrolysis reactor, a water jacket, a burner device and a steam-gas mixture pipeline.

After the geometry of the PP tubular reactor is constructed, it is necessary to set the type of material for each unit of the device for simulating the thermal process. In this case, the following types of materials were selected for each part of the PP tubular reactor: water – for filling the water jacket, ordinary water was selected; structural steel – for the walls of the reactor, the burner device, the steam–gas mixture pipeline and for the chimney pipe; wood was selected as biomass (humidity – 10%). For the designation of steam gas, the type of material steam is selected, and for the designation of smoke, the type of material smoke is selected.

To accurately simulate the thermal process, polygons were created for each type of material. In Fig. 5 you can see the polygons created for each part and each type of material that are isolated from each other.

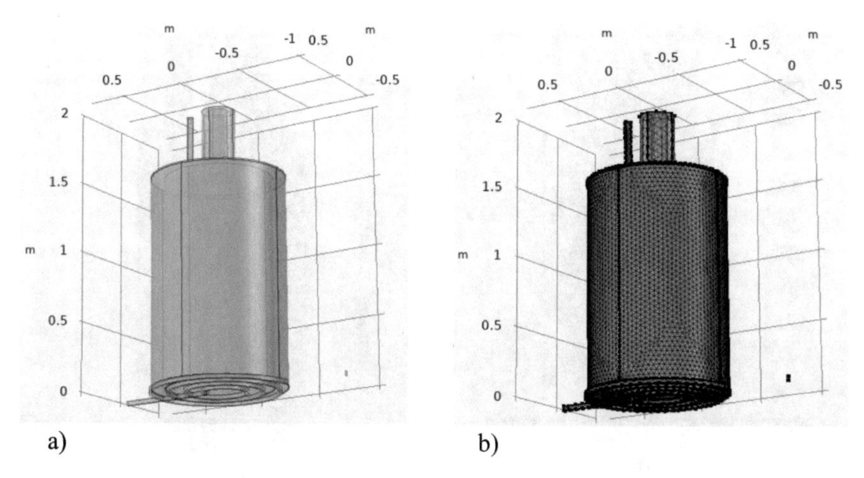

Fig. 4. 3D model of the PP tubular reactor (a – normal view, b – mesh view).

The principle of operation of the PP tubular reactor is based on heating hydrocarbon materials to high temperatures ($t = 20 \div 700$°C), which leads to their thermolysis – the decay of molecules [14–16].

In our case, a tubular PP reactor with a water jacket is used, which serves to recover heat from the pyrolysis reactor to the water that is in the water jacket and the heated water ($t = 18 \div 80$°C) can be used in heat supply systems. And also the water jacket helps to control the temperature inside the reactor and prevents overheating [12, 17].

Comsol Multiphysics and many of its add-on modules include embedded materials libraries, which are databases with materials and related material properties.

These libraries contain up to 24 different properties for each material and a set of 10328 substances with more than 84,000 sets of data on properties. Almost all physical properties are given as functions of temperature.

Piecewise set polynomial functions of temperature T are usually used to describe temperature dependences, so the input parameter for such functions can be the temperature calculated as a result of heat transfer modeling.

Libraries contain data on the properties of various modifications or different states of the same substance, for example, in different ranges of state parameters.

Thus, Comsol Multiphysics establishes the physical properties of materials based on extensive built-in libraries that contain detailed data about a variety of materials and their properties.

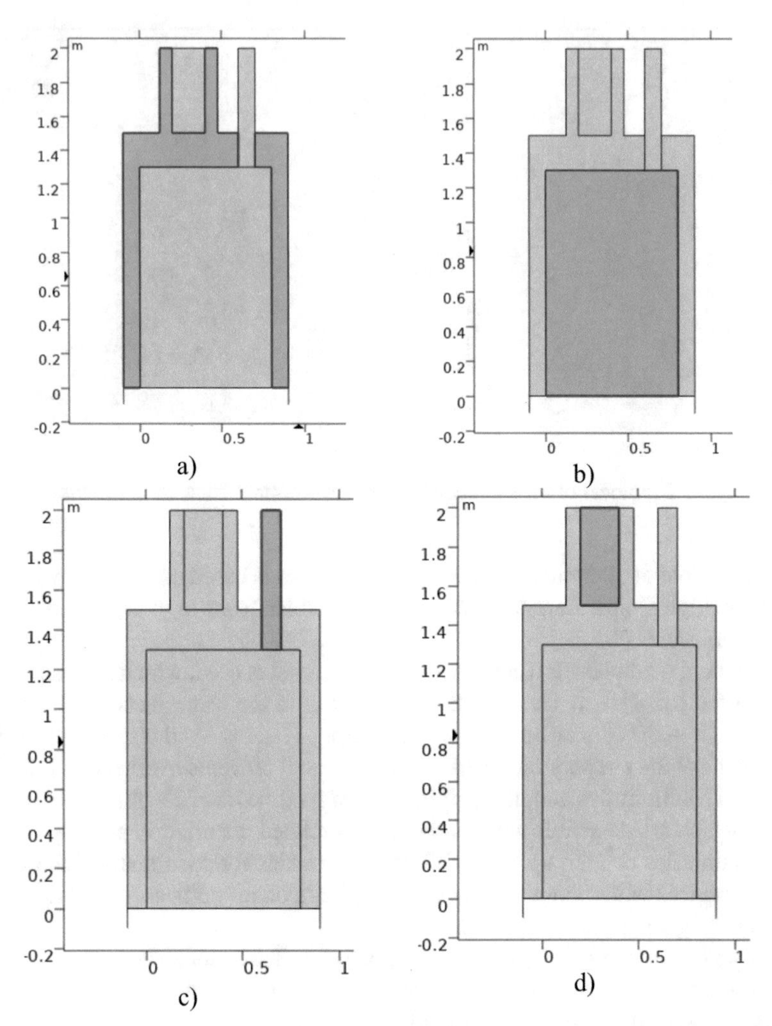

Fig. 5. Polygons of each type of material (a – the landfill of the water jacket, which consists of the water material, b – the landfill of the pyrolysis reactor, which consists of the dry wood material, c – the landfill of the steam–gas mixture pipeline, which consists of the steam material, d – the landfill of the chimney pipe, which consists of the smoke material).

3 Results

To conduct a scientific experiment of the decomposition of biomass at high temperatures in the absence of oxygen. Birch wood was selected as biomass, which has the following thermophysical properties: the thermal conductivity of birch wood is 0.142 W/ (m*S), the average density of birch wood is about 650 kg/m^3 at a relative humidity of 12–15% [18].

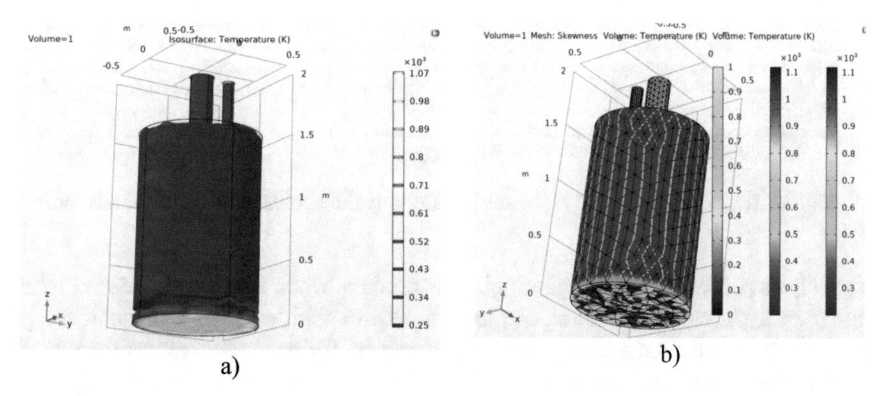

Fig. 6. Changes in the tubular reactor after 1 min of heating: a) isosurface view; b) mesh view.

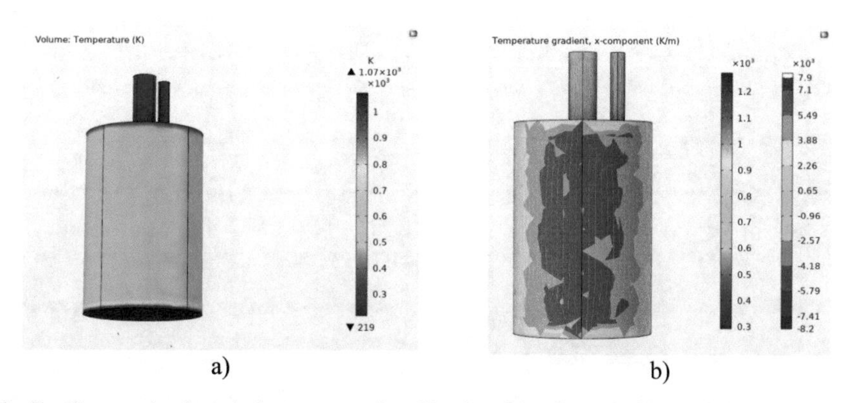

Fig. 7. Changes in the tubular reactor after 60 min of heating: a) 3D model; b) view of the isosurface.

To study the physics of heat transfer in solids, initial values, temperatures, boundary heat sources and other physical parameters have been established that will contribute to heat transfer in solids.

As a result of the study, the following results were obtained, which visualize the process of heat exchange inside a tubular PP reactor with different duration of the decomposition process of plant biomass.

In graph (Fig. 8.), you can see the temperature change along the length of the tubular reactor. The temperature decreases as the arc length increases, which may be due to

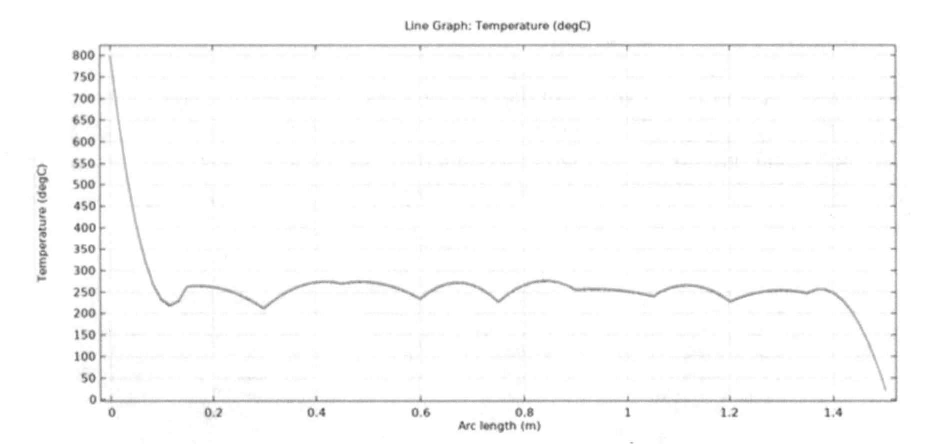

Fig. 8. Temperature change along the y axis along the vertical of the tubular reactor.

the pyrolysis process. At the beginning of the process, when the temperature is highest, long-chain hydrocarbons decompose into molecules with a lower molecular weight. This leads to the formation of products that can be further processed in the existing infrastructure. Over time, as we move along the length of the reactor, the temperature decreases.

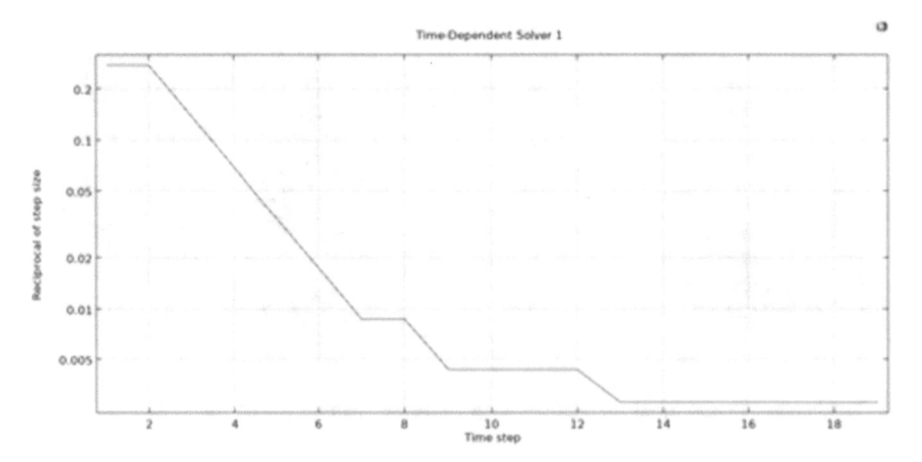

Fig. 9. Changes in reaction rate over time

The graph (Fig. 9) shows the results of pyrolysis in a tubular reactor. The X-axis is designated as the "Time step", and the Y-axis is designated as the "Reaction rate of container species". The line starts with a high reaction rate of container species at time step 0 and steadily decreases to time step 18. The reaction rate of container species decreases from about 0.2 at time step 0 to about 0.005 at time step 18. This suggests that over time, the reaction rate of container species in the pyrolysis process decreases.

It is important to note that specific process conditions and parameters, such as pressure, flow rate and mixture composition, can significantly affect the pyrolysis process and, consequently, the temperature profile in a tubular reactor. Additional data may be required for a more accurate analysis.

4 Discussion

The obtained results of a scientific experiment of the decomposition of biomass at high temperatures in the absence of oxygen on the developed model of a tubular PP reactor using Comsol Multiphysics software (shown in Fig. 6. And Fig. 7.), represent a three-dimensional representation of the decomposition process of biomass. The model for the study of heat transfer in the pyrolysis process includes the thermal conductivity, density and dynamic viscosity of the loaded biomass in the pyrolysis reactor. These thermophysical properties play a key role in determining the pyrolysis efficiency and can be accurately modeled using Comsol Multiphysics software.

Figure 6 on the right shows the temperature scale in kelvins. It indicates that the model is used to visualize the temperature distribution inside the reactor during the pyrolysis process. The mesh view allows you to visualize the geometry of the reactor and visualize the distribution of temperature or other parameters inside the reactor.

The models obtained by studying the process using software Comsol Multiphysics in the form of isosurfaces (Fig. 6. (a) and Fig. 7. (b)) are a three-dimensional analogue of an isoline, that is, a surface representing points with a constant value (in this case temperature) in some part of the space of a tubular reactor. In other words, it is a set of the level of a continuous function whose domain of definition is a three-dimensional space. Isosurfaces are derived using computer graphics and are used as visualization methods in computational thermal dynamics.

5 Conclusions

Comsol Multiphysics is one of the best software that contributes to the visual implementation of scientific research and will provide an opportunity to better understand and optimize the pyrolysis process. The model of a PP tubular reactor in a simplified geometry allows you to significantly reduce the amount of computational work during modeling, and also allows you to reduce the calculation time by 72%. This model can be used to study various aspects of pyrolysis with different biomass feed of different composition and type, including heat transfer, decomposition reactions and gas flow, and can also be used to analyze decomposition reactions that occur during pyrolysis. This is important for understanding how biomass is converted into biofuels during pyrolysis. In general, the PP tubular reactor model is a tool for researching and optimizing the pyrolysis process.

This model will be used for further research of a pyrolysis plant with a regenerative heat exchanger.

References

1. Sattorov, B., Davlonov, K., Toshmamatov, B., Arziev, B.: Increasing energy efficiency combined device solar dryer-water heater with heat accumulator. BIO Web Conf. **71**, 024 (2023)
2. Toshmamatov, B., Kodirov, I., Davlonov, K.: Determination of the energy efficiency of a flat reflector solar air heating collector with a heat accumulator. E3S Web Conf. **402**, 05010 (2023)
3. Uzakov, G., Mamatkulova, S., Ergashev, S.: Modeling of heat exchange processes in a condenser of a pyrolysis bioenergy plant. BIO Web Conf. **71**, 02021 (2023)
4. COMSOL Multiphysics Simulation Software: Understand, Predict, and Optimize Real-World Designs, Devices, and Processes with Simulation, https://www.comsol.ru/comsol-multiphysic. Last accessed 20 Oct 2023
5. Simulating Heat Transfer in Layered Materials. https://www.comsol.com/support/learning-center/article/41591. Last accessed 20 Oct 2023
6. Kaczor, Z., Buliński, Z., Werle, S.: Modelling approaches to waste biomass pyrolysis: a review. Renew. Energy **159**, 427–443 (2020)
7. Uzakov, G., Mamatkulova, S., Ergashev, S.: Thermal mode of the condenser of a pyrolysis bioenergy plant with recuperation of secondary thermal energy. E3S Web Conf. **411**, 01021 (2023)
8. Toshmamatov, B., Davlonov, K.: Recycling of municipal solid waste using solar energy. IOP Conf. Ser. Mater. Sci. Eng. **1030**(1) (2021)
9. Akhralov, S., Yusupov, R., Egamberdiev, K.: Geoinformation technologies and methods of mathematical modeling in hydrogeological research. InterCarto. InterGIS **26**, 240–252 (2020)
10. Mitrofanov, A.V., Mizonov, V.E., Vasilevich, S.V., Malko, M.V.: Experiments and computational research of biomass pyrolysis in a cylindrical reactor. In: Proceedings of CIS Higher Education Institutions and Power Engineering Associations. Energetika **64**(1), 51–64 (2021)
11. Toshmamatov, B., Davlonov, K., Rakhmatov, O., Toshboev, A., Rakhmatov, A.: Modeling of thermal processes in a solar installation for thermal processing of municipal solid waste. AIP Conf. Proc. **2612**, 050027 (2023)
12. Davlonov, K., Toshboev, A., Sultanov, S.: The main heat-technical parameters of solar greenhouses in the southern climate of Uzbekistan. E3S Web Conf. **411**, 01038 (2023)
13. Comsol Multiphysics User's Guide Comsol AB, Stockholm, Sweden Version 5.0 (2014)
14. Tubular pyrolysis reactor: description, operating principle and application. https://rospatenta.ru/i/trubcatyi-reaktor-piroliza-opisanie-princip-raboty-i-primenenie. Last accessed 20 Oct 2023
15. Mamatkulova, S.G., Uzakov, G.N.: Modeling and calculation of the thermal balance of a pyrolysis plant for the production of alternative fuels from biomass. In: IOP Conference Series: Earth and Environmental Science, 1070 (2022)
16. Mamatkulova, S.: Improvement of the GIS map of the potential of biomass for the development of bioenergy production in the Republic of Uzbekistan. In: AIP Conference Proceedings, 2432 (2022)
17. Laudert, D.: Mythos Baum, 57–63. BLV, München (2009)
18. Toshmamatov, B., Shomuratova, S., Safarova, S.: Improving the energy efficiency of a solar air heater with heat accumulator using flat reflectors. E3S Web Conf. **411**, 01026 (2023)

An Overview on Data Augmentation for Machine Learning

Svetlana Volkova(✉)

Vologda State University, 15 Lenin Street, 160000 Vologda, Russia
malysheva.svetlana.s@gmail.com

Abstract. The effective utilization of data augmentation stands as a strategic imperative in the domains of industrial enterprises and economics, offering a potent means to enhance the convergence and performance of machine learning models. Data augmentation, as a methodological approach, plays a crucial role in generating additional data from existing datasets, thereby expanding the dataset's scope. This approach proves especially valuable when confronted with limited dataset sizes, a common challenge in these domains. The essence of model generalization in industrial and economic contexts not only demands an expansion in data volume but also necessitates diversification to effectively capture the intricacies of real-world scenarios. Data augmentation, by generating diverse instances from existing datasets, plays a crucial role in addressing these imperatives, mitigating overfitting risks, and bolstering model robustness. Data augmentation is a versatile technique applicable to various data types prevalent in business and industry, including numerical, categorical, and textual data, as well as images, audio, and video. This article provides a systematic exploration of data augmentation techniques across different data categories. It begins with an introduction to the problem statement, followed by sections dedicated to data augmentation for tabular data, text data, and image data. The article concludes by underscoring the importance of combining diverse strategies and evaluating their impact on machine learning models.

The applicability of data augmentation extends across a spectrum of areas within these domains, including supply chain optimization, financial forecasting, market analysis, and operational efficiency enhancement. Ultimately, the incorporation of diverse data augmentation strategies emerges as a significant factor in bolstering the stability, reliability, and generalization capacity of machine learning models.

Keywords: Machine Learning · Deep Learning · Data Augmentation · Overfitting

1 Introduction

The relentless expansion of machine learning and intelligent data analysis finds applications in diverse domains, including industrial enterprises, economics, and management. Machine learning techniques play a pivotal role in solving complex tasks within these sectors, ranging from computer vision [1, 2] to natural language processing [3, 4] and medical diagnostics [5, 6].

A. Gibadullin (Ed.): DITEM 2023, LNNS 942, pp. 143–154, 2024.
https://doi.org/10.1007/978-3-031-55349-3_12

The stochastic gradient algorithm (SGD) [7, 8] serves as a foundational method for training machine learning models in these applied contexts. Stochastic gradient enables models to learn from data and improve their ability to solve various machine learning tasks crucial to industrial enterprises and economic analyses, including classification, regression, and more. The iterative adjustment of model parameters based on errors in training data contributes to incremental improvements in model quality, augmenting its generalization capability.

Despite the clear advantages of the stochastic gradient algorithm, such as computational efficiency and versatility, this method has a set of limitations and drawbacks. The stochastic gradient can sometimes suffer from the issue of overfitting, where the model becomes overly tailored to the training data and struggles to generalize well to new data. Overfitting results in the model developing overly complex and overly specific rules, leading to poor generalization and low performance on new examples. To address this problem, various methods are applied, including regularization, the addition of dropout layers, or the use of data augmentation to create a more diverse set of training examples.

This article focuses on data augmentation as a means to improve the convergence of machine learning models.

Enhancing the convergence of machine learning models encompasses the utilization of data augmentation as one of its instrumental strategies. Data augmentation stands as a methodological approach for the generation of supplementary data from pre-existing datasets. This method involves the deliberate expansion of the dataset's scope.

Data augmentation is applied in situations when the dataset's scale is restricted. Achieving superior model generalization necessitates not only a larger volume of data but also a heightened diversity in the data between iterations. Data augmentation functions as the catalyst for introducing novel data variants, thereby amplifying the model's capacity for generalization while concurrently mitigating the risk of overfitting.

Applicable across various data types, including numerical, categorical, textual, images, audio, and video, data augmentation emerges as a valuable approach for enhancing the efficiency of machine learning model training in industrial and economic applications.

Many researchers have made efforts to categorize augmentation methods. Nevertheless, it is most commonly observed that methods and models are primarily examined within the context of a single modality, like text, images, or signals. For example, in [9, 10], methods for augmenting textual data are discussed. Various image augmentation techniques are outlined in [11–13]. Methods for augmenting tabular data are presented in [14, 15].

This article provides a systematic overview of data augmentation techniques for various types of data. The article is organized as follows: in the first section, we outline the problem statement, the second section discusses data augmentation techniques for tabular data, the third section is dedicated to text data augmentation, the fourth covers image augmentation, and, finally, we conclude the whole paper.

2 Tabular Data Augmentation

During the augmentation of tabular data, our aim is to artificially increase the size and diversity of data represented in the form of a table. In this data representation, each column typically corresponds to a specific feature or attribute, and each row signifies an individual object, observation, or data point. This section provides a classification of several well-established techniques for augmenting tabular data, such as noise adding, data shifting, synthetic minority oversampling, and random missing values adding.

2.1 Random Noise Adding

The method involves adding random values to existing data, which helps introduce variability and diversity into the original dataset. Let us now explore several methodologies for adding random noise.

i. Gaussian Noise
   For each (or a randomly chosen) element within the table, the addition of a random value drawn from a normal distribution with a mean of zero and a specified variance is possible. This process facilitates the generation of minor, stochastic deviations from the initial values.
ii. Uniformly Distributed Random Noise
   This approach involves adding random values from a uniform distribution within a specified range. As a consequence, this can lead to more substantial and varied alterations within the dataset.
iii. Noise with Feature Selection
   Random noise can be introduced by choosing a random feature from the dataset and modifying its value by a certain random magnitude. This process contributes to the diversification of each individual feature.

It is imperative to note that when incorporating random noise into the dataset, a thorough understanding of the data's context and intrinsic characteristics is essential. The indiscriminate or excessive addition of noise can potentially distort the data and result in unfavorable consequences.

2.2 Data Shifting

An additional technique for augmentation of tabular data is data shifting, which involves modifying their values by applying an offset or multiplier. Let us now delve into specific instances of data shifting.

i. Multiplication by a Coefficient
   In this approach, the values of each feature are subject to multiplication by a randomly chosen coefficient, which can be either slightly greater or slightly smaller than one. This method has the potential to alter the amplitude of feature values.
ii. Shifting Numerical Values
   Random values within a predefined range can be added to each feature's value. This procedure can lead to alterations in the mean values of the features and introduce variability into the dataset.

iii. Shifting Temporal Data

In order to infuse diversity into temporal data, temporal timestamps can be displaced either forward or backward by a random duration of time.

iv. Shifting Categorical Values

In the case of categorical features, the process of shifting involves substituting their values with alternative values sourced from either the same feature or neighboring categories. For instance, one can substitute the value "red" in the categorical feature "color" with a randomly selected value from other categories, such as "crimson" and "pink." This practice creates a novel data variant that upholds the overarching category of "color" while introducing diversity in the specific shade of color.

Similar to the addition of random noise, it is crucial to consider the context and specific characteristics of the features when shifting data. Furthermore, semantic considerations are important to avoid shifts that contradict the nature or interpretation of the data. For example, if an object is described as belonging to the "cat" class, which has four legs, shifting the feature to 3 or 5 no longer accurately represents a "cat."

2.3 SMOTE (Synthetic Minority Oversampling Technique)

SMOTE (Synthetic Minority Oversampling Technique) [15] is a methodology employed in the realm of machine learning to address the issue of imbalanced datasets. It achieves this by generating synthetic samples for the minority class (with insufficient representation), based on existing samples. The core principle of SMOTE involves the application of interpolation techniques between the nearest neighbors of existing data points to construct novel samples.

The SMOTE process can be delineated as follows:

i. Selection of a sample from the minority class.
ii. Determination of k nearest neighbors from the same class.
iii. Random selection of one the neighbor.
iv. Generation of a new synthetic sample by adding a weighted difference between the selected neighbor and the original sample. The weight is a random number within the range [0, 1].
v. Iteration through the preceding four steps as necessary to generate the desired quantity of synthetic samples.

Fundamentally, SMOTE's core concept revolves around finding the nearest neighbor from the same class for a given object, constructing a line connecting these two objects, and the subsequent selection of a random point along this line. It is important to note that the selection can be carried out not only along the line connecting the chosen points but also independently for each coordinate.

To elucidate the data augmentation process within the context of a classification problem that hinges on two features and makes use of SMOTE, let us contemplate the ensuing example. The initial dataset is presented in the Table 1.

Table 1. Initial dataset of SMOTE elucidation.

Feature 1	Feature 2	Class label
0.23	1.12	2
0.54	0.89	2
0.65	0.65	1
0.22	0.22	1
1.11	1.23	2
2.2	2.25	0
3.2	1.1	0

Let us examine the first sample from class “1”, characterized by features (0.65, 0.65), and its nearest neighbor from the same class. Within our dataset, the nearest neighbor corresponds to the object with features (0.22, 0.22). The line connecting these two objects assumes the form $y = x$. To create a new sample, we can select any point along the $y = x$ line, situated between (0.22, 0.22) and (0.65, 0.65). For the purposes of this illustration, we shall select the point (0.3, 0.3). Subsequent to the application of SMOTE, the dataset is expanded by the addition of a novel synthetic sample to class “1,” and it takes the form presented in the Table 2.

Table 2. Dataset after application of SMOTE (class “1” expanding).

Feature 1	Feature 2	Class label
0.23	1.12	2
0.54	0.89	2
0.65	0.65	1
0.22	0.22	1
1.11	1.23	2
2.2	2.25	0
3.2	1.1	0
0.3	0.3	1

Then look at the last sample from class “0”, characterized by features (3.2, 1.1), and its nearest neighbor from the same class. Within our dataset, the nearest neighbor from the same class corresponds to the object with features (2.2, 2.25). The line connecting these two objects assumes the form $y = -1.15x + 4.78$. To create a new sample, we can select any point along the $y = -1.15x + 4.78$ line, situated between (2.2, 2.25) and (3.2, 1.1). We shall select the point (2.6, 1.79). After the application of SMOTE, the dataset is expanded by the addition of a novel synthetic sample to class “0,” and it takes the form presented in the Table 3.

Table 3. Dataset after application of SMOTE (class "0" expanding).

Feature 1	Feature 2	Class label
0.23	1.12	2
0.54	0.89	2
0.65	0.65	1
0.22	0.22	1
1.11	1.23	2
2.2	2.25	0
3.2	1.1	0
0.3	0.3	1
2.6	1.79	1

Consequently, the utilization of SMOTE yields the benefits of rectifying sample imbalances, augmenting dataset dimensions, and bolstering the model's ability to generalize minority class data more effectively. However, in the application of SMOTE, as with other augmentation methodologies, it is important to exercise control over the generation process to prevent the creation of excessively noisy or redundant samples, which could adversely impact model performance.

2.4 Random Missing Values Adding

The random missing values adding stands as another technique for augmenting tabular data. This strategy involves the random elimination of data points within certain features or observations, thereby facilitating the training of the model on datasets with incomplete records. The addition of random missing values can aid the model with the capacity to learn from data that exhibits gaps, effectively preparing it to handle analogous situations in real-world scenarios.

However, it is imperative to exercise prudence when engaging in data removal processes. Missed values can contain crucial information, and their elimination may introduce distortions in results and lead to incorrect conclusions. Furthermore, it is essential to remain of the fact that the inclusion of random missing values has the capacity to perturb the data distribution, potentially shifting it towards a different configuration.

Despite these considerations, the incorporation of random missing values can enhance the model's capability to handle absent data, thereby improving its robustness.

As a consequence, when an augmentation approach is selected, it is prudent to conduct a comprehensive analysis of task-specific intricacies, data modalities, and the intended objectives. Combining various augmentation strategies and conducting a meticulous assessment of their impact on the results can significantly enhance the model's generalization capacity and stability.

3 Text Data Augmentation

In textual data augmentation, the objective is to artificially expand the size and diversity of data represented in the form of text, encompassing words, phrases, sentences, or their vectorized representations. Given that textual data is predominantly harnessed in natural language processing (NLP) tasks, text data augmentation is primarily utilized to enhance the effectiveness of NLP models in addressing tasks such as machine translation, text classification, sentiment analysis, and text generation. This section presents a categorization of some well-established techniques for textual data augmentation, such as synonym replacement, vector representation replacement, text translation to another language and noise adding.

3.1 Synonym Replacement

In the synonym replacement [16, 17], the procedure involves the random substitution of a variable number of words with their synonyms, all the while preserving the semantic content of the sentence. It's worth noting that this substitution isn't limited to individual words; phrases and expressions can also be replaced with synonymous alternatives of similar meaning.

The process of synonym replacement can be implemented using a synonym dictionary, or by utilizing NLP models trained to retrieve synonyms or synonymous phrases. For instance, consider the original sentence *"Artificial* intelligence *conquers the world"*, which can be substituted with *"Artificial intelligence prevails over the world"*. In this example, the word "conquers" was substituted with "prevails over", while the semantic meaning of the sentence remained unaltered.

It is imperative to recognize that in the process of synonym replacement, due consideration must be given to the context and the intricacies of word combinations and sentence structures. Substituting a word such as "artificial" in the previous example with "unnatural" could potentially result in text distortion and misinterpretation.

3.2 Close Vector Representations Replacement

The technology of replacing similar vector representations [18, 19] is akin to the synonym replacement technique, with the key distinction being that, in this case, word replacements are sought not from a synonym dictionary but based on the distances between their vector representations.

Various models (e.g., Word2Vec) can serve as vector representation extractors, representing words as numerical vectors. The process of substituting a random number of words with their vector representations unfolds as follows:

i. Conversion of each word in the text into the corresponding vector using a pretrained word vector representation model.
ii. Calculation of the similarity between the vectors of words in the sentence and the vectors of words from the entire vector space. Various metrics can be employed for this purpose, for example Euclidean or Cosine distance.

iii. Selection of a random number of words in the original text and their replacement with words having the closest computed similarity.

Substituting words with similar vector representations can be particularly advantageous in tasks where there is a need to enhance the models' generalization capabilities and make them more resilient to variations in text.

3.3 Translation of Text (or Its Portions) to Another Language and Back

Translating a portion of text into another language and back [20, 21], using machine translation, allows for the creation of variations in phrases and expressions and can lead to the replacement of certain constructs. During this process, the semantic meaning of the text or sentence is retained.

The augmentation process unfolds as follows: Firstly, the original text is translated into the target language using machine translation tools. Many services offer APIs to automate the translation process, which can be beneficial when working with large volumes of data. Subsequently, the translated text in the target language is translated back into the original language. Various machine translation tools may be employed for this reverse translation, and it's important to note that these tools may not match perfectly during both the forward and backward translation.

Translating to another language and back facilitates introducing variations into textual data, expanding the vocabulary, and creating diversity in expressions and phrases. For example, if we translate the phrase "*Artificial intelligence prevails over the world*" from English to Latin using Google Translate, we get the phrase "*Intelligentia artificialis praevalet mundo*", which, when translated back, sounds like "*Artificial intelligence is taking over the world*".

It is not necessary to limit the translation to a single direction. For example, the original (e.g., English, "*Artificial* intelligence *prevails over the world*") phrase can be translated into Language #1 (e.g., Spanish, "*La inteligencia artificial prevalece en el mundo*"), then the result of that translation can be translated into Language #2 (e.g., Serbian, "*U svetu preovladava veštačka inteligencija*"), and only then back into the original language (e.g. English, "*The world is dominated by artificial intelligence*"). This can result in even greater data variability.

3.4 Noise Adding

The last techniques discussed within the scope of this textual data augmentation classification are grouped under the category of "Noise Adding".

This category encompasses the introduction of random noise into the text, such as typos and the insertion of random or special characters. Additionally, grammatical errors can be inserted as noise, assisting the model in learning to handle them, and enhancing its resilience [22, 23]. For example, it is possible to add mistakes into the phrase "*Artificial intelligence prevails over the world*" and modify it to "*Artificial intelligence prevail over the world*" or "*Artifisial intelligence prevails over the world*". Also, additional spaces or other punctuation marks can be added, e.g., "*Artificial intelligence prevails over the world!*". It can lead to processing greater variability data.

Furthermore, to create new variations of phrases and expressions, the word order can be altered, and for diversification in text length and structure, some random words or symbols can be removed (e.g., *"Artificial intelligence prevails the world"*).

4 Image Data Augmentation

Let us now turn our attention to the discussion of image data augmentation techniques. Image augmentation is employed to enhance the effectiveness of computer vision models. The principal objective of image data augmentation is to artificially enrich and diversify the dataset, empowering computer vision models to effectively handle a broader spectrum of real-world scenarios. By introducing variations in the training data, these techniques enable models to become more resilient, adaptable, and capable of generalizing well beyond the limited scope of their initial training data. This is of paramount importance in computer vision applications such as object detection, image classification, facial recognition, and more, where robustness to variations in lighting, perspective, and object appearance is critical. In this section, techniques, that have become essential tools in the arsenal of computer vision practitioners, were enumerated and elucidated.

4.1 Horizontal/Vertical Mirroring (Flip)

In horizontal or vertical image flipping [12], the image is mirrored horizontally or vertically, creating a mirrored or upside-down version of the original image.

Horizontal flipping involves pixel reversal relative to a vertical axis running through the center of the image. This means that any object or element that was originally positioned on the left side of the image will now appear on the right after this transformation, and conversely, those on the right side will be shifted to the left. This simple yet effective operation can effectively simulate variations in object orientation and is useful for enhancing the training data for computer vision models.

Similarly, in vertical flipping, the reversal is conducted relative to a horizontal axis passing through the image's central point. Consequently, objects that were initially located at the bottom of the image will be relocated to the top, and those originally at the top will now be at the bottom post-transformation. This technique is valuable for simulating changes in the perspective or orientation of objects within the image.

Reflection operations, whether horizontal or vertical, are particularly useful in situations where objects in the image are symmetric or lack a clear orientation. Computer vision models after train images flipping become more adept at recognizing and generalizing object features, irrespective of their spatial arrangement or orientation.

4.2 Random Image Cropping (Random Crop)

The procedure of randomly excising a segment of an image is sometimes denoted as "random cropping" [24, 25]. This method entails the extraction of a random rectangular region from the source image and subsequently assigning it to the source image.

In the course of the cropping process, coordinates are randomly chosen to ascertain both the position and dimensions of the cropped region. Additionally, it is possible to impose constraints pertaining to the upper and lower thresholds for points displacement with respect to the image boundaries, as well as for the maximum and minimum dimensions of the cropped region.

The addition of random cropping enhances the robustness to changes in scale and shifts of objects relative to the image.

4.3 Brightness and Contrast Adjustment

The manipulation of image brightness and contrast affords the capacity to generate diverse lighting scenarios [12].

Brightness modifications are realized through the implementation of linear operations on pixel intensities. The most straightforward approach entails the multiplication of all pixel values by a scaling factor, denoted as "k". When $k > 1$, brightness increases, and when $k < 1$, it decreases proportionally. For instance, a multiplication of 1.15 elevates the image's brightness by 15%, whereas a multiplication of 0.85 decreases it by 15%.

In the context of contrast, its adjustment involves the scaling of the distribution of pixel values. For example, employing histogram equalization to achieve a uniform distribution of pixel values can result in an augmentation of image contrast.

It is important to underscore that uncontrolled or excessively intense adjustments of brightness and contrast should be avoided, as they may lead to information loss or significant distortion of the image.

4.4 Random Noise Adding

The random noise adding can be employed to induce variations in the texture and fine details of an image [26, 27]. Various types of noise, such as Gaussian noise and salt-and-pepper noise (which involves the random addition of white and black pixels), can be applied to individual pixels in the image. Alternatively, one can apply a random multiplication factor to each pixel, and color information within the pixels can also be perturbed through random adjustments of color channel values.

Each of these transformations can, to varying degrees, prove advantageous in enhancing a model's capacity for generalization and bolstering its robustness against noise encountered in real-world scenarios.

Examples of data augmentation on the image of the digit "7" from MNIST [28] dataset can be observed in Fig. 1

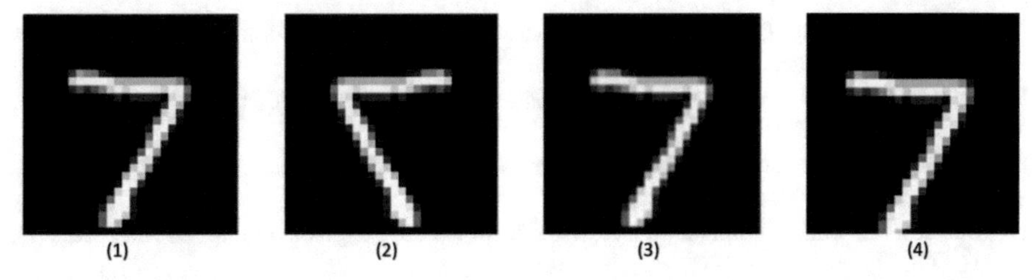

Fig. 1. Examples of data augmentation on the image of the digit "7" from MNIST, including the original image (1), reflection or flip (2), blurring (3), and random cropping (4).

5 Conclusion

Data augmentation encompasses the process of generating supplementary data through the application of diverse transformations to pre-existing datasets. This methodology is particularly advantageous in situations where dataset limitations necessitate an expansion in both volume and diversity to bolster a model's generalization prowess while mitigating the perils of overfitting.

Data augmentation can be harnessed across a spectrum of data types, spanning numerical, categorical, textual, visual, auditory, and audio-visual domains. It is imperative to acknowledge that the efficacy of data augmentation is intricately linked to the specific task, data modalities, and targeted outcomes. The present study systematically categorizes augmentation methodologies for tabular data, text, and imagery, underscoring the significance of amalgamating diverse strategies and appraising their impact on machine learning models.

Data augmentation stands as a substantial contributor to the enhancement of model performance within the domains of natural language processing, computer vision, and beyond, ultimately elevating their capacity for generalization and training stability.

References

1. Sebe, N.: Machine Learning in Computer Vision. Springer Science & Business Media 29 (2005)
2. Volkova, S.S., Bogdanov, A.S.: A deep learning approach to face swap detection. Int. J. Open Inf. Technol. **9**(10), 16–20 (2021)
3. Khan, W.: A survey on the state-of-the-art machine learning models in the context of NLP. Kuwait J. Sci. **43**(4) (2016)
4. Jiang, H., et al.: Smart: Robust and efficient fine-tuning for pre-trained natural language models through principled regularized optimization. arXiv preprint arXiv, 1911.03437 (2019)
5. Bhavsar, K.A., et al.: Medical diagnosis using machine learning: a statistical review. Comput. Mater. Contin. **67**(1), 107–125 (2021)
6. Vu, T., Nguyen, D.Q., Nguyen, A.: A label attention model for ICD coding from clinical text. arXiv preprint arXiv, 2007.06351 (2020)
7. Hecht-Nielsen, R.: Theory of the Backpropagation Neural Network. Neural Networks for Perception. Academic Press, pp. 65–93 (1992)
8. Bottou, L.: Online Algorithms and Stochastic Approximations. Online Learning and Neural Networks. Cambridge University Press (1998)

9. Bayer, M., Kaufhold, M.A., Reuter, C.: A survey on data augmentation for text classification. ACM Comput. Surv. **55**(7), 1–39 (2022)
10. Feng, S.Y., et al.: A survey of data augmentation approaches for NLP. arXiv preprint arXiv, 2105.03075 (2021)
11. Shorten, C., Khoshgoftaar, T.M.: A survey on image data augmentation for deep learning. J. Big Data **6**(1), 1–48 (2019)
12. Buslaev, A., et al.: Albumentations: fast and flexible image augmentations. Information **11**(2), 125 (2020)
13. Yang, S., et al.: Image data augmentation for deep learning: a survey. arXiv preprint arXiv, 2204.08610 (2022)
14. Onishi, S., Meguro, S.: Rethinking data augmentation for tabular data in deep learning. arXiv preprint arXiv, 2305.10308 (2023)
15. Chawla, N.V., et al.: SMOTE: synthetic minority over-sampling technique. J. Artif. Intell. Res. **16**, 321–357 (2002)
16. Zhang, X., Zhao, J., LeCun, Y.: Character-level convolutional networks for text classification. In: Advances in Neural Information Processing Systems 28 (2015)
17. Dai, X., Adel, H.: An analysis of simple data augmentation for named entity recognition. arXiv preprint arXiv, 2010.11683 (2020)
18. Nie, Y., et al.: Named entity recognition for social media texts with semantic augmentation. arXiv preprint arXiv, 2010.15458 (2020)
19. Wan, Z., Wan, X., Wang, W.: Improving grammatical error correction with data augmentation by editing latent representation. In: Proceedings of the 28th International Conference on Computational Linguistics (2020)
20. Sennrich, R., Haddow, B., Birch, A.: Improving neural machine translation models with monolingual data. arXiv preprint arXiv, 1511.06709 (2015)
21. Li, Y., et al.: A diverse data augmentation strategy for low-resource neural machine translation. Information **11**(5), 255 (2020)
22. Feng, S.Y., et al.: Genaug: Data augmentation for finetuning text generators. arXiv preprint arXiv, 2010.01794 (2020)
23. Bishop, C.M.: Training with noise is equivalent to Tikhonov regularization. Neural Comput. **7**(1), 108–116 (1995)
24. Takahashi, R., Matsubara, T., Uehara, K.: Data augmentation using random image cropping and patching for deep CNNs. IEEE Trans. Circuits Syst. Video Technol. **30**(9), 2917–2931 (2019)
25. Duong, H.T., Nguyen-Thi, T.A.: A review: preprocessing techniques and data augmentation for sentiment analysis. Comput. Social Netw. **8**(1), 1–16 (2021)
26. Moreno-Barea, F.J., et al.: Forward noise adjustment scheme for data augmentation. In: IEEE Symposium Series on Computational Intelligence (SSCI) (2018)
27. Bae, H.J., et al.: A Perlin noise-based augmentation strategy for deep learning with small data samples of HRCT images. Sci. Rep. **8**(1), 17687 (2018)
28. LeCun, Y., et al.: Gradient-based learning applied to document recognition. Proc. IEEE **86**(11), 2278–2324 (1998)

Keyboard Handwriting as a Factor of User Authentication

Stanislav Zateev(✉) and Alexander Shelupanov

Tomsk State University of Control Systems and Radioelectronics, Tomsk, Russian Federation
zateev.stanislav@gmail.com

Abstract. Currently, information technologies are used in all spheres of our life. Huge amounts of personal and corporate information are stored digitally, the disclosure of which can bring serious problems to an individual or a company, may entail reputational losses or financial damage. In this regard, there is a problem of protecting this information, ensuring its security. Currently, one of the most common methods of information protection is means of identification and authentication. According to modern information security standards, information about the dynamics of the user's work with the keyboard can be used as authentication data. The development of an continual authentication system based on the dynamics of the user's work with the keyboard is divided into 2 subtasks: the formation of a dataset from raw data by identifying features and training the classifier on the prepared data. At the stage of forming suitable data sets, the source data is divided into windows, and a set of statistical features for analysis is allocated in each window. At the stage of using the classifier, the optimal parameters of the classifier for training are selected. The resulting model is able to accurately determine the legitimacy of the user by a set of characteristics of the dynamics of his work with the keyboard. Moreover, it provides opportunities for further research on the possibility of building an continual authentication system based on the analysis of keyboard handwriting.

Keywords: Authentication · Keystroke dynamics · Features of keystrokes and their combinations · Decision Trees · Random Forests

1 Introduction

One of the most effective and frequently used means of information protection in the modern world is the means of user authentication. Such measures are a convenient means for delimiting user access to system resources. However, it should be recognized that the means of authentication are not sufficiently reliable.

As you know, authentication means (and methods) are divided into three types according to the type of factors used:

- Authentication means that use what the user knows (passwords, passphrases, PIN codes, answers to key questions, etc. are used as a factor).
- Authentication means using what the user has (USB tokens, passports, magnetic cards, RFID identifiers, etc. are used as a factor).

A. Gibadullin (Ed.): DITEM 2023, LNNS 942, pp. 155–167, 2024.
https://doi.org/10.1007/978-3-031-55349-3_13

- Authentication means using biometric characteristics of the user (fingerprints, retinal scan, etc. are used as a factor).

Obviously, the first two types of factors are relatively easily compromised: a password or other secret knowledge can be stolen, transferred to an attacker, often such systems can be circumvented by selecting a password using the simplest search algorithms, similarly, a real identifier can be stolen, lost, compromised in another way.

Unlike the first two types, authentication means based on the biometric factor are more reliable: they cannot be stolen or lost, and they are also very difficult to fake. However, in order for a biometric characteristic to be used as an authentication factor, the user must be completely determined by a unique set of parameter values of this characteristic.

Authentication means using the biometric factor can be divided into two groups: physiological (using static characteristics: fingerprint or retina scan) and behavioral (using dynamic characteristics: voice or gait). The use of physiological characteristics increases the reliability of the authentication tool compared to other types of factors, but they are also compromised [1]. Behavioral characteristics are more reliable, but they can change over the course of life or depending on the physical condition of the user, so they are less stable. Also, biometric authentication means usually require expensive equipment necessary for information retrieval.

An important disadvantage of authentication means is their "milestone" nature: the inability to detect the fact of a user change in the system.

Therefore, I especially want to highlight such a biometric characteristic as keyboard handwriting – a set of dynamic characteristics of the user's work on the keyboard. Keyboard handwriting is suitable for use as an authentication factor, which is confirmed by a large number of researches and works devoted to this topic [2–9]. Unlike most biometric authentication means, it does not require additional equipment other than a conventional keyboard. It is also very important that the keyboard handwriting can be used both for the standard authentication procedure [2–4] and for the formation of an continual authentication tool [5, 6]: because the user spends a considerable amount of time working with the keyboard during the session, you can build a software product that allows you to check the legitimacy of the user throughout the entire keyboard session.

In the process of studying the topic, already implemented algorithms for using keyboard handwriting as an authentication factor were considered. For example, one of the methodologies [5] consists in splitting the data into time intervals, allocating statistical characteristics of the sample within the intervals and training a model using the K-nearest neighbor method on these data. Another methodology [7] consists in identifying features in the event stream (the duration of keystrokes and the intervals between keys) and training a model using the support vector method on the received data. The third methodology [8] is based on the formation of a data set consisting of the retention durations of bigraphs (combinations of 2 keys) that occur more than a fixed number of times, and the training of a Gaussian mixture model on it. Another methodology [2] involves splitting all the keys on the keyboard into 32 groups and calculating 5 features for each group, then a neural network is trained on the resulting data set.

This article is devoted to the research of the construction of a system that allows user identification.

2 Materials and Methods of Research

2.1 Dataset

In the process of research, specially collected data were used for it.

A solution implemented in Windows Forms format was used to collect data. The program records the process (each event) of typing by the user. The following parameters are recorded for each event: user ID, session ID (users were recorded in several sessions), key ID, event type (pressing or releasing a key) and timestamp. Thus, 26 users were recorded. The obtained data underwent a preprocessing procedure (removal of garbage taps, removal of repeated events, etc.) for further comfortable use.

Table 1 presents statistics on the received data after preprocessing. As you can see, a user with the ID 2 stands out from the total mass of users, he has twice as many records as the average number of records of other users. However, within the framework of a multiclass classification, with a fairly equal distribution of records from other users, this will not upset the balance of the sample.

Table 1. User statistics.

User ID	Number of sessions	Number of entries	User ID	Number of sessions	Number of entries
2	25	510 664	18	9	234 689
3	6	188 463	19	6	182 515
4	4	188 063	20	31	211 512
6	4	191 009	21	7	236 307
7	5	184 795	23	5	228 161
8	4	164 261	29	10	255 472
9	3	185 459	32	12	250 625
10	5	189 021	35	18	281 819
12	11	252 394	36	6	246 696
13	10	243 792	37	5	250 522
14	14	225 277	38	9	254 273
16	15	252 418	39	8	252 173
17	8	273 610	40	27	280 831

The next step is to determine the principle of generating a dataset and identifying features to enable the use of this data in a machine learning model.

The data set is formed as follows. The event stream, the format of which is described at the beginning of the section, is divided into fixed-size intervals, as shown in Fig. 1. Then, within each interval, the values of the features are calculated, which will be described further. A vector of feature values describing this interval is collected from the calculated values. The set of such vectors will be the final dataset.

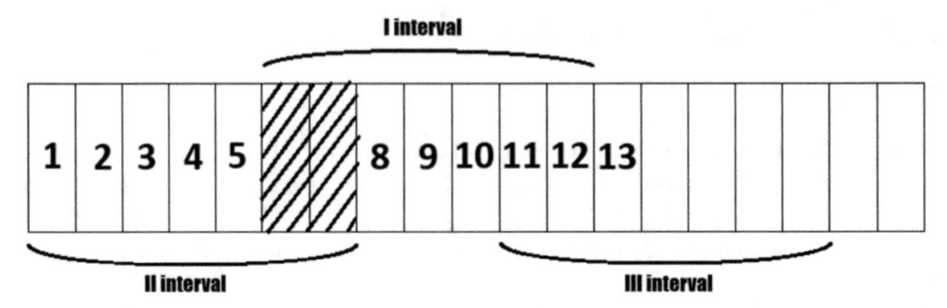

Fig. 1. The principle of dividing the flow into intervals.

This principle of data generation determines two key parameters, the change of which will have a significant impact on the quality of the model: the size of the interval (the larger the interval, the more informative it is, the longer the conversion of raw data into a vector of feature values will take place) and the percentage of overlap between intervals (allows neighboring intervals to take into account each other's context). 2000 and 1000 events are taken as the base values of the interval size. The percentage of overlap will vary from 0 to 75 according to the scheme shown in Fig. 2.

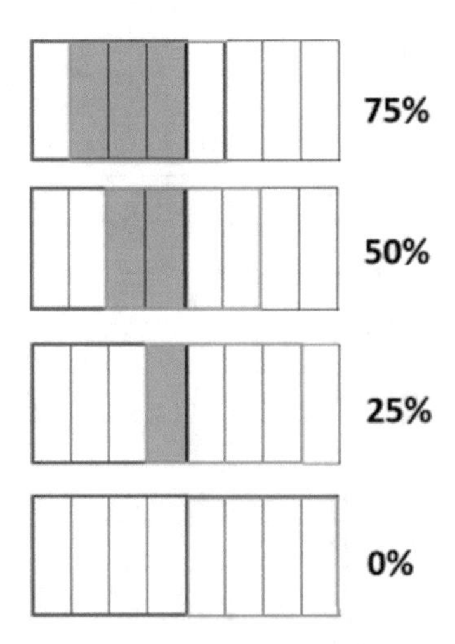

Fig. 2. Options for splitting into intervals.

An important factor is also the set of distinguished features. At the initial stages of the work, common features characterizing the user were distinguished (typing speed, percentage of errors, degree of arrhythmicity of typing, etc.). However, such features are not enough to form a feature space, also these features are not very informative, therefore, their use is unnecessary.

Therefore, the final feature space is formed from the parameters of pressing individual keys and their combinations [9–11]. However, it should be understood that when using the parameters of all keys and their combinations (of course, the length of the combination is limited), the feature space will grow to an incredible size, but at the same time 99% of the features will be absolutely uninformative, which will greatly reduce the quality of the machine learning model. Researches [11] show that a limited number of keyboard shortcuts are sufficient to achieve acceptable accuracy. The parameters of the most frequently used keys and their combinations were taken as features in the final model. 30 single keys, 70 bigrams (combinations of two keys) and 20 trigrams (combinations of 3 keys) were selected. Figure 3 shows the statistics of the use of keys in the collected data (left) and the statistics of the use of letters in Russian (right). As you can see, the sample turned out to be quite representative.

	frequence	key (russian)	english equivalent
32	335427	(Space)	(Space)
74	218761	о	o
84	198616	е	e
66	173003	и	i
89	156900	н	n
70	154680	а	a
78	145945	т	t
8	125885	(BackSpace)	(BackSpace)
67	117753	с	s
72	108924	р	r
68	90931	в	v
75	85003	л	l
86	71216	м	m
160	69714	(LShift)	(LShift)
191	62940	.	.

equivalent	frequences
o	о - 11.35%
e	е - 8.93%
a	а - 8.23%
n	н - 6.71%
i	и - 6.48%
t	т - 6.17%
s	с - 5.22%
l	л - 4.95%
v	в - 4.47%
r	р - 4.17%
k	к - 3.35%
d	д - 2.97%
m	м - 2.93%
u	у - 2.86%

Fig. 3. Comparison of letter usage frequencies.

Also at the first stages of the work, combinations of 4 and 5 keys and words (constructions from space to space) were used. However, as practice has shown, due to the fact that such constructions are rare, in most intervals it is impossible to calculate the value of certain parameters. Because these values cannot be calculated, they are filled in based on the values of the same parameters, but for intervals in which they can be calculated (these cells are filled with the median value for this parameter for the user being researched). Because of this, such parameters for the machine learning model are characteristic, which is not true. This leads to incorrect operation of the model as a whole.

The parameter of pressing a single key is one – the duration of holding this key. The choice of parameters for pressing keyboard shortcuts is slightly more complicated: it is necessary to select such a set of parameters that would fully describe the combination, but at the same time would not be redundant. Figure 4 shows the principle of determining the parameters of a trigram. Feature 1 is the time interval between pressing the first key and pressing the second; feature 2 is the time interval between pressing the first key and releasing the second, etc.

Bigrams are characterized by features from the first to the third.

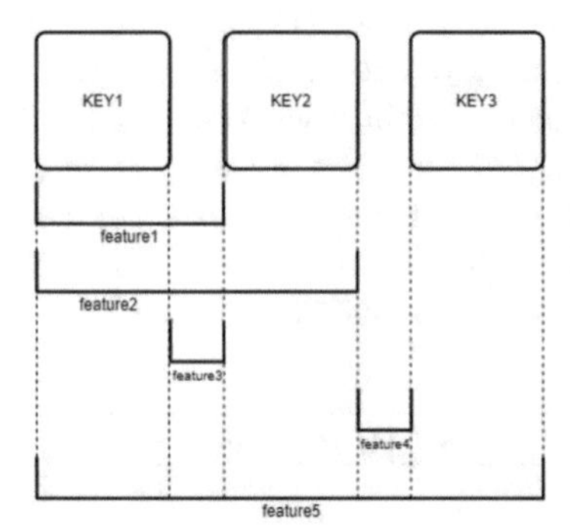

Fig. 4. Features of pressing trigrams.

It is also important to note that keystrokes and shortcuts within the same interval may occur more than once. Consequently, it is not the parameters themselves described above that are taken as final features, but the statistical characteristics of the parameter samples within the interval (sample mean, standard deviation and skewness coefficient).

Thus, the generated dataset is a set of vectors. The dimension of the vector (feature) space is 1020. The number of entries in the dataset varies around 10000 (depends on the interval parameters when splitting).

2.2 Classifier

Obviously, the further implementation of the product requires the use of machine learning methods, a classifier that should determine the user based on the selected features. The choice fell on the classifiers "Decision Tree" and "Random forest".

Decision trees are a logical decision-making scheme consisting of a hierarchically distributed system of questions. Each question is a partition of a certain component of the vector feature space by a certain value. The construction of the system of questions is based on the principle of greedy maximization of information gain – the question is asked in such a way that the answer to it brings as much information as possible.

However, the use of separate decision trees usually leads to overfitting of the model and poor quality of predictions. Therefore, it would be most appropriate to use a classifier that combines the results of several trees – a forest.

A random forest is a collection of a certain number of trees. The forest splits the training data randomly (including the features are split) and distributes it among the trees, each tree is trained on its own data and the classifier averages the final result.

Decision trees and random forests were chosen for the following reason. The initial hypothesis was that if the selected features can describe a certain user, then among all the features it is possible to single out a set that would be characteristic for this user, that is, it took values mainly within a certain limited interval. This assumption fits perfectly with the principle of how trees work.

Here it is worth going back a bit, to the definition of the most commonly used keys and combinations. As mentioned earlier, in the first versions of the product, tetragrams, pentagrams and words were taken into account, among others. But the probability of their meeting in the interval is extremely low, this leads to the fact that in most intervals it is impossible to calculate the value of certain parameters. Because these values cannot be calculated, they are filled in based on the values of the same parameters, but for intervals in which they can be calculated (these cells are filled with the median value for this parameter from the user being researched). But there are much more indefinite intervals than definite ones, this leads to the fact that for most intervals of a certain user, a certain parameter takes the same value, i.e. a certain value of a certain parameter becomes characteristic for a given user. Figure 5 shows an example of such data in an existing dataset.

89|70 2 mean	89|70 3 mean	class
84.0	151.0	2
84.0	151.0	2
84.0	151.0	2
84.0	151.0	2
84.0	151.0	2
...	...	...
84.0	151.0	2
84.0	151.0	2
84.0	151.0	2
84.0	151.0	2
143.0	47.0	2

Fig. 5. Data illustrating the revealed pattern.

Naturally, clipping by such a parameter will reduce entropy more than clipping by any of the remaining ones. Obviously, in a real situation, such a division is incorrect, since the sample on the basis of which the classifier makes a decision is not representative. Figure 6 shows an example of how the program converts data. As you can see in the figure, there is a data set of 8 positions, 3 of them are known, the remaining 5 positions are filled with the median value (in this case 49). This data set is part of some conditional dataset (Fig. 7 – parameter x1). With this configuration of the dataset, the tree (1) will be the optimal solution for the decision tree model. However, in reality this parameter (x1) is not a characteristic for object 1 ($y = 1$), so the tree (2) will be the more correct tree in this case. At the same time, the model showed a high result even on the test sample because the splitting of the initial data into training and test data occurred after filling in the empty cells of the intervals for which it is impossible to calculate the corresponding parameters.

That is, the distribution of values in the test and training samples are coordinated so that the test sample obeys the training logic. That is why it is very important to choose the right number of structures under research – not to reduce too many features and not to create an excess of uninformative features.

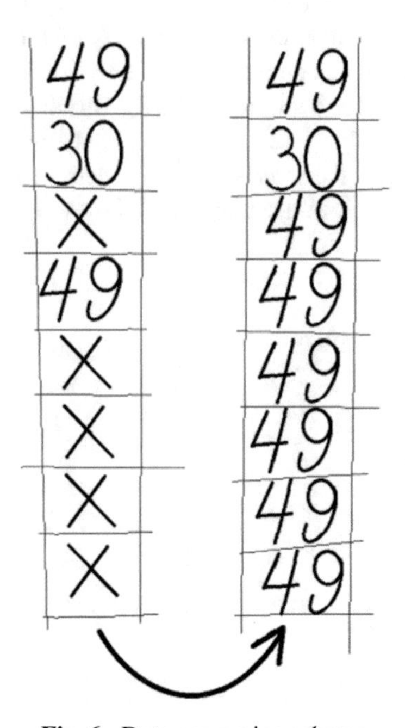

Fig. 6. Data conversion scheme.

Trees and forests have a set of key parameters that significantly affect the quality of the model: the depth of the tree (limiting the depth will not allow the model to retrain, it will not allow taking into account statistically unimportant examples), the minimum size of the sheet and the minimum size of the sheet to split (similarly to the depth, limiting the size of the sheet will not allow taking into account statistically garbage examples), the criterion of informativeness (a criterion for implementing the principle of greedy maximization of information gain), the number of trees (only for forests).

Both decision trees and random forests were used for the key search. The following parameters were used for the trees:

- Criterion – Shannon entropy, Gini index;
- Maximum depth – from 5 to 10;
- The minimum sheet size is from 20 to 50 (in increments of 10);
- The minimum size of the split node is from 20 to 50 (in increments of 10).

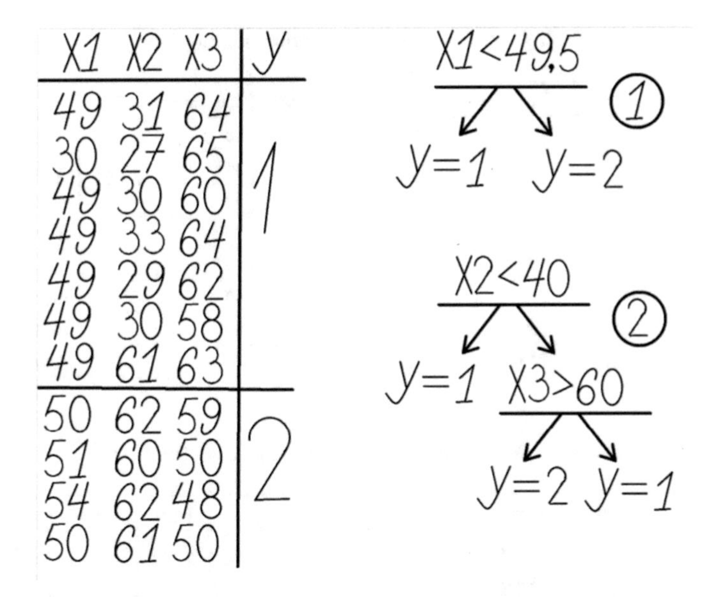

Fig. 7. Tree construction scheme.

As parameters for forests, we used:

- Criterion – Shannon entropy, Gini criterion;
- Number of trees – from 60 to 80 (in increments of 10);
- Maximum depth – from 5 to 9;
- The minimum sheet size is from 10 to 30 (in increments of 10);
- The minimum size of the split node is from 10 to 30 (in increments of 10).

The training was carried out using cross–validation. The number of samples is 4.

3 Results

The results of the models are summarized in Tables 2 and 3. In the columns size and step, the parameters for splitting into intervals are specified: size – the size of the interval (1000 and 2000 events), step – the step that determines the percentage of overlap of the intervals (0%, 25%, 50% and 75%). The accuracy column contains the accuracy of the model prediction on the test sample. The remaining columns contain the optimal parameters of the model for prediction on the given data.

Confidence intervals for prediction accuracy values were also calculated. The parameters of confidence intervals are calculated using the following formulas: interval radius (delt) – according to formula 1, lower and upper bounds (min and max) – according to formula 2. The results of calculations of these parameters are presented in Tables 4 (predictions using decision trees) and 5 (predictions using random forests) (Table 5).

$$delt = z \cdot \sqrt{\frac{accuracy \cdot (1 - accuracy)}{n}}, \tag{1}$$

Table 2. Parameters of models during training (trees).

cv	accuracy	criterion	max_depth	min_samples_leaf	min_samples_split	size	step
4	0.8590	entropy	7	20	40	2000	2000
4	0.8795	entropy	8	20	30	2000	1500
4	0.8835	entropy	8	20	50	2000	1000
4	0.9195	entropy	9	20	20	2000	500
4	0.8446	entropy	9	20	30	1000	1000
4	0.8764	entropy	9	20	40	1000	750
4	0.8877	entropy	9	20	30	1000	500
4	0.9230	entropy	10	20	40	1000	250

Table 3. Parameters of models during training (forests).

cv	accuracy	n_esimators	criterion	max_depth	min_samples_leaf	min_samples_split	size	step
4	0.9878	70	entropy	8	10	10	2000	2000
4	0.9933	80	entropy	8	20	10	2000	1500
4	0.9960	70	entropy	9	10	10	2000	1000
4	0.9966	80	entropy	9	10	10	2000	500
4	0.9946	80	entropy	8	10	10	1000	1000
4	0.9959	80	entropy	9	10	30	1000	750
4	0.9978	60	entropy	9	10	30	1000	500
4	0.9965	60	entropy	9	10	10	1000	250

where z is the critical value from the Gaussian distribution (2.58 with 99% reliability), n is the dataset size.

$$border = pred \pm delt, \tag{2}$$

where pred is the predicted value.

Table 4. Parameters of confidence intervals (trees).

size	step	DF_size	delt	min	max
2000	2000	2730	0.01718	0.841863	0.876228
2000	1500	3619	0.01396	0.865536	0.893460
2000	1000	5360	0.01130	0.872246	0.894854
2000	500	10613	0.00681	0.912684	0.926312
1000	1000	5573	0.01252	0.832043	0.857087
1000	750	7404	0.00987	0.866565	0.886299
1000	500	11032	0.00776	0.879911	0.895425
1000	250	21847	0.00465	0.918371	0.927677

Table 5. Parameters of confidence intervals (forests).

size	step	DF_size	delt	min	max
2000	2000	2730	0.00542	0.982369	0.993214
2000	1500	3619	0.00350	0.989808	0.995802
2000	1000	5360	0.00221	0.993831	0.998255
2000	500	10613	0.00146	0.995111	0.998037
1000	1000	5573	0.00254	0.992024	0.997106
1000	750	7404	0.00191	0.993994	0.997822
1000	500	11032	0.00115	0.996653	0.998953
1000	250	21847	0.00103	0.995507	0.997559

4 Discussion

As can be understood from the above results, using decision trees and random forests, it is possible to achieve high classification results. The maximum accuracy value of the decision trees is 0.923. In this case, the confidence interval is [0.918; 0.927]. This accuracy is already satisfactory for using the model as an authentication tool, especially in continual authentication tools, the principle of operation of which will allow averaging the predictions of the model over a certain period.The accuracy of the trees in this case coincides with the results of the analogs [12].

The accuracy of predicting random forests is even higher. The maximum achieved accuracy is 0.9978 with a confidence interval [0.997; 0.999]. In general, all the assembled models give a guaranteed accuracy of 98–99%, which makes these models suitable even for authentication without averaging readings.

Table 6 compares the model based on the use of random forests with the analogues described in the introduction.

It is also important to understand that the set of parameters of machine learning models for the research was very limited. As you can see in Table 3, the optimal value of the number of trees and the depth of the tree often takes a boundary value, which may indicate that there is a more optimal value beyond the boundaries of the values used. The set of values of the partitioning parameters into intervals was also very limited.

Thus, the obtained models showed high results sufficient to guarantee their suitability for use as an authentication tool. At the same time, the ceiling of quality values was not reached.

Table 6. Comparison with analogues.

Reference	Preprocessing	Features	Method	Accuracy
–	Splitting the event stream into time windows	Statistical characteristics of samples of calculated parameters within intervals	Random forest	0.98
[2]	Splitting keys into 32 groups	5 features for each group	Neural network	0.95
[5]	Splitting the event stream into time windows	Statistical characteristics of the sample within the intervals	k-nearest neighbors method	0.96
[7]	–	Pressing time and intervals between keystrokes	Method of support vectors	0.03 (FRR)
[8]	Selection of bigrams that occur more often than a certain number of times	Duration of pressing the most common bigrams	Model of Gaussian mixtures	0.94

5 Conclusion

The work done has once again proved the suitability of keyboard handwriting for use as an authentication factor. Models implementing the classification of sets of keyboard handwriting parameters by users were configured and trained, which showed high accuracy results.

The available models are an excellent basis for further formation of the continual user authentication system. Moreover, keyboard handwriting, which is a biometric characteristic, is a hard-to-compromise factor, an authentication tool based on the analysis of keyboard handwriting does not require any equipment for data retrieval other than a standard keyboard. At the same time, as researches have shown, the reliability of the authentication tool collected in this way will be very high.

Acknowledgements. This research was funded by the Ministry of Science and Higher Education of Russia, Government Order for 2023–2025, project no. FEWM-2023-0015 (TUSUR).

References

1. Cao, K., Jain, A.K.: Hacking mobile phones using 2D printed fingerprints. Technical report (2016)
2. Kim, J., Kang, P.: Recurrent neural network-based user authentication for freely typed keystroke data. arXiv preprint arXiv:1806.06190 (2018)

3. Kang, P., Cho, S.: Keystroke dynamics-based user authentication using long and free text strings from various input devices. Inf. Sci. **308**, 72–93 (2015)
4. Bailey, K.O., Okolica, J.S., Peterson, G.L.: User identification and authentication using multi-modal behavioral biometrics. Comput. Secur. **43**, 77–89 (2014)
5. Monaco, J.V., et al.: Developing a keystroke biometric system for continual authentication of computer users. In: Intelligence and Security Informatics Conference (EISIC), pp. 210–216 (2012)
6. Bakelman, N., et al.: Continual keystroke biometric authentication on short bursts of keyboard input. In: Proceedings of Student-Faculty Research Day. CSIS, Pace University (2012)
7. Yu, E., Cho, S.: GA-SVM wrapper approach for feature subset selection in keystroke dynamics identity verification. In: 2003 Proceedings of the International Joint Conference on Neural Networks, pp. 2253–2257 (2003)
8. Ceker, H., Upadhyaya, S.: Enhanced recognition of keystroke dynamics using Gaussian mixture models. In: MILCOM 2015–2015 IEEE Military Communications Conference, pp. 1305–1310 (2015)
9. The, P.S., Teoh, A.B.J., Yue, S.: A survey of keystroke dynamics biometrics. Sci. World J. (2013)
10. Tappert, C., Cha, S., Villani, M., and Zack, R.: A keystroke biometric system for long-text input. Int. J. Inf. Secur. Priv. 32–60 (2010)
11. Al, Solami, E., Boyd, C., Clark, A., Ahmed, I.: User-representative feature selection for the keystroke dynamics. In: Proceedings of 2011 5th International Conference on Network and System Security, pp. 229–233 (2011)
12. Alsultan, A., Warwick, K.: Keystroke dynamics authentication: a survey of free-text. Int. J. Comput. Sci, Issues **10**, 1 (2013)

Development of an Algorithm for Constructing Complex Fractal Images Using the L-Systems Method for Textile and Building Materials Production Enterprises

Z. E. Ibrohimova(✉) and Sh. A. Omonkulova

Samarkand branch of the Tashkent University of Information Technologies named after Muhammad al-Khorezmi, Samarkand, Uzbekistan

Zuli117@mail.ru

Abstract. This article covers some elements of the theory of fractals, methods, and algorithms for constructing fractals. The analysis of studies on the L-systems among the fractals construction methods, algorithms for constructing some fractal images, and the obtained results are presented below. Using the L-systems method, a geometric model for creating new complex fractal images using geometric substitutions from dragon-shaped, tree-shaped, and spiral fractals was built. And the results were obtained using the developed software. In this article, the software tool developed on the basis of the L-systems method, the algorithm for creating complex fractal structures using geometric substitutions can be effectively used in the textile industry, in the design of building materials, and in the technologies of printing patterns on gas and carpets. The use of the developed algorithm and software tool helps to create a wide range of pattern types in a short period of time in textile and building materials development enterprises, reduces costs and increases the production volume and demand for the product.

Keywords: Fractal · L-systems · Dragon Fractal · Tree Fractal · Spiral Fractal

1 Introduction

The development and development of technologies for the use of fractal geometric shapes in the field of design of the textile industry remains one of the most important issues in the world. Considering the creation of fractal images, the application of fractals in the field of textiles is divided into two parts. First, fractal images are used in the design of textile patterns, and secondly, the properties of woven fabric such as permeability analysis, fabric defect detection, fabric surface texture analysis, etc. are analyzed based on the theory of fractals. The use of fractal images provides new creative ideas for designing textile patterns. Fractal theory is a powerful tool for solving complex problems in the textile industry.

Today, the research on the mathematical aspects of fractal theory and the methods of describing natural processes and phenomena using the ideas of fractal theory is under

A. Gibadullin (Ed.): DITEM 2023, LNNS 942, pp. 168–179, 2024.
https://doi.org/10.1007/978-3-031-55349-3_14

special attention. Specifically, fractal theory, computer graphics methods, and systems are widely used in constructing mathematical equations of fractal shapes [1–5].

Hungarian botanist Aristide Lindenmayer proposed a grammar system for the biological model of plant growth patterns in 1968 [6, 7]. Lindenmayer systems, or L-systems for short, were initially developed as a mathematical theory of plant development. This study presents examples of an algorithm for creating self-similar fractal patterns using the L-systems method.

On the one hand, the regularity of L-systems is rewriting. It is a technique for detecting complex objects by sequentially permuting parts of a simple initial object using a collection of rewriting rules [8].

On the other hand, recursion is a key component of fractal geometry, which is the repeated application of a rule to successive results. Therefore, the concept of rewriting L-systems is an effective mechanism for applying recursion to achieve complex fractals [9–11].

The main applications of fractals in textile engineering include textile image design, textile weave design, textile pattern design, etc. For example, patterns created by a complex iterative method are compositions of a set. And make them look great by applying different functions and color schemes. They can be used to design gas pattern images. Fractal patterns based on the method of IFS and L-systems are created based on the property of strict self-similarity, and they can be used in the design of the textile industry [11].

In textile pattern design, we can see the application of fractal image mainly in clothing pattern design, fashion pattern design and decorative gauze pattern design, etc. [10]. Li Chen et al. automatically generated an M-set using Visual Basic (VB). Successful images are obtained by editing images using Photoshop and applied to knitted items. Using the VB language, Baoqi Wang et al. studied the visualization of J-set and higher-order J based on time-lapse algorithm. Also studying Julia paintings used in Jacquard designs. J-sets are depicted in single-layer jacquard gauze using a weave as shown in Fig. 1.

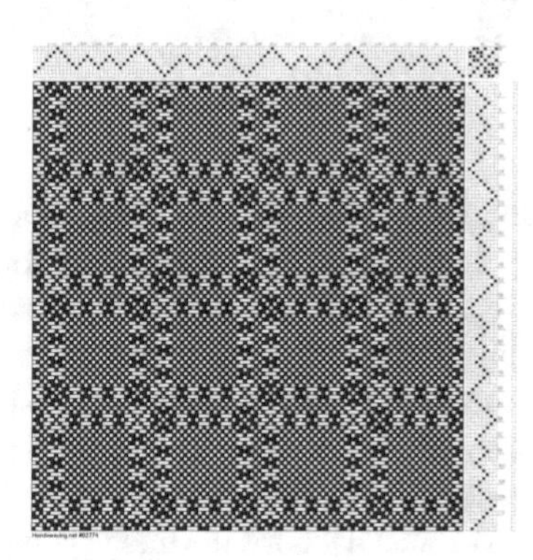

Fig. 1. Fabric weaving design based on L-system.

In textiles, weft design is often carried out by the L-system with self-similarity in mind. Based on the characteristics of a simple weaving structure, Kejun Cen et al. used a fractal with a self-similarity level of 4 using the L-system creation method. Figure 2 shows examples of fractal gauze woven on an electronic jacquard machine.

Fig. 2. A fractal pattern used in L-system textile pattern design.

Jacquard fabric is a complex and simple woven fabric, based on which there are more than 24 different woven threads [12; p. 132]. Jacquard fabrics are also made of colorful patterns. Jacquard fabrics have the following characteristics: product durability, color brightness, washing resistance, easy cleaning, beautiful appearance, etc. There are several types of Jacquard fabrics: Jacquard - satin, Jacquard - silk, Jacquard - satin, Jacquard - knitted, Jacquard - stretch, etc. (see Fig. 3).

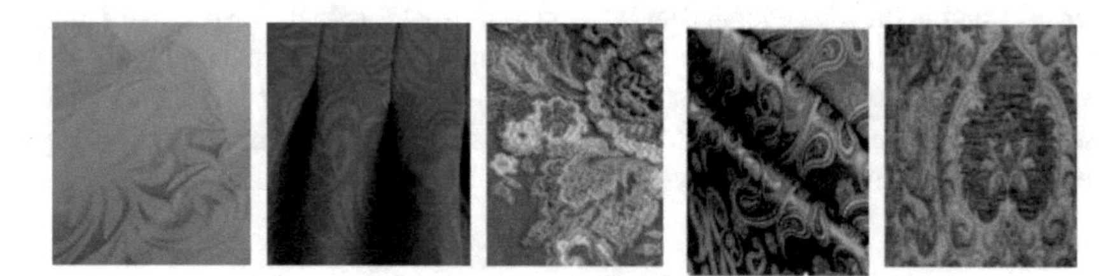

Fig. 3. Jacquard fabrics.

The pattern design of Jacquard embroidery is always one of the main components that scholars are devoted to study. In recent years, many forms of pattern have been shown to be used in the design of knitted garments. For example, Cai Yanyan studied how to systematically develop fractal images and wrote a book entitled "Research and Application of Fractal Geometry in Design of Apparel Models". Fractal graphics are used in the design of clothes and give a table of unique effects. Wang Shuyin's book "Learning and Applying Fractal Theory to Costume Design" provides instructions on how to learn the fundamentals of fractal theory, the fractal graphics design algorithm, and how to use design research to create clothing patterns based on previous research. He then used Photoshop, Corel Draw, and rendered fractal graphics to obtain design patterns and applied these design patterns to the design of clothing and clothing accessories. Young analyzed the use of fractal patterns in digital jet printing of suits in French-style costume design and the use of fractal images in Custom Made [6].

2 Mathematical Model

L-systems method

The Lindenmayer system was proposed in 1968 by the Hungarian botanist Aristide Lindenmayer to study the development of simple multicellular organisms. Later, he extended complex branching structures (trees, florals), to model formal languages, biological breeding models, fractals, etc. [12].

To implement the L-systems the so-called Turtle (turtle algorithm) graphics setups are used as part systems. In this case, the turtle point moves on the screen with discrete steps drawing its track, as in the rule, or, if necessary, moves without drawing points [13].

Assume that the executors of the set of commands are turtles. The turtle moves along the plain. Let x and y be the coordinates of the turtle's initial position, and let α be the angle defining the direction taken by the turtle. Imagine a turtle has a memory. Given the coordinates of the turtle's initial location x0, y0 and the direction of movement α0, as well as the value of the step h, the turtle moves according to the "forward" command and turns to the angle b according to the right or left command [11, 12], [13].

- Make the turtle able to execute the following commands:
- "F" - leaving a trail h step forward in direction α;
- "f" - h step forward in α direction without leaving a trace;
- "+" - turn to the right at an angle β (clockwise);
- "−" - turn to the left at an angle β (clockwise);
- "[" - (x, y, α) to save the initial state to memory;
- "]" - remember the last saved position (x, y, α).
- As an example, we observe the process of constructing Koch triad curves. This curve was studied by the Swedish scientist Helge von Koch in 1904 [12].

The *L*-system consists of a word formed through an iterative process and an algorithm for its graphical interpretation. Legitimately, an expression consists of the command symbols F, b, $+$, $-$, [,] and is formed as follows. The starting word simultaneously generates rules parallel to the axiom $axiom = W^0 = W^0\ (F, b, +, -, [,])$. Then the process is repeated to obtain a sequence of words W^0, W^1, W^2, ... with a complex internal structure of symbols.

For example, using the axiom *F* and the rule F+F-F-F+F, we can generate the W^0, W^1, W^2, ... sequence:

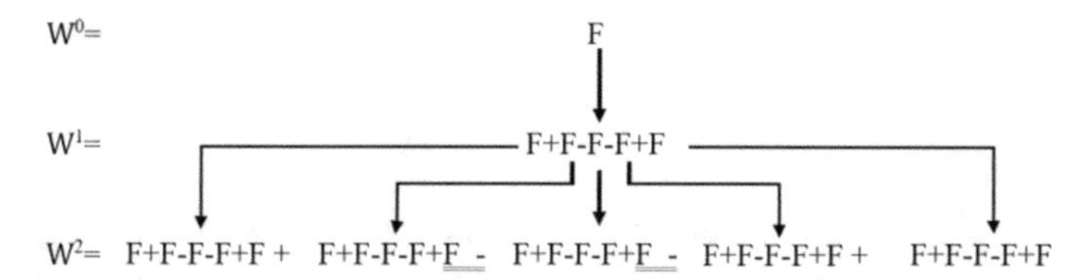

See the work of the turtle in the example of the words W^0, W^1, W^2 taken in the example above (where $\theta = \pi/2$, the length scale in all images is different) (Fig. 4):

The following obscurity arises if fractals are constructed using only one generating rule. We cannot change the reading direction of the rule at some stages, fi.e. we can

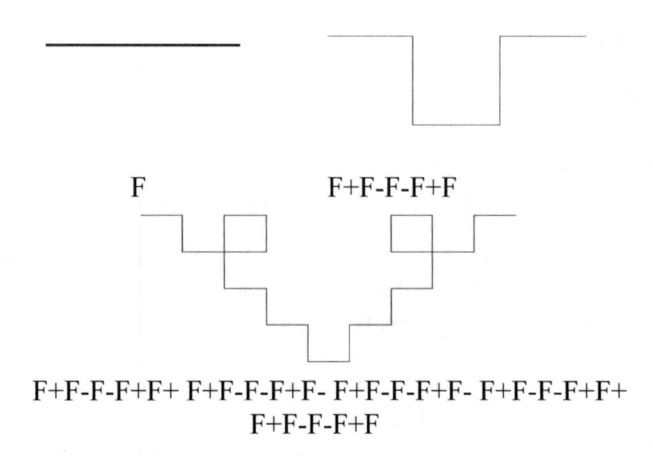

Fig. 4. *L*-system and turtle graph implementation

read right-to-left instead of left-to-right. Without solving this problem, it is impossible to obtain L-systems for various classes of curves (Peano curve, Harter-Hateway dragon, Hagerty tile, etc.). For example, to construct a fractal called the "Harter-Hathaway Dragon", it is necessary to be able to change the reading direction of the generation rule shown in the Fig. 5. The curve on the left is used as a primer or axiom. The rule of descent, in this case is to pull the initiator forward first and then back. Such a scheme does not fit into the scope of L-systems, which use only one producer rule. This problem can be solved by entering two different commands to move forward, say X and Y. Suppose the turtle interprets X and Y as the same, i.e. one step forward. With these two letters, the generative rule for the dragon can be written as follows:

axiom = X,
newx = X +Y+,
newy = -X-Y.

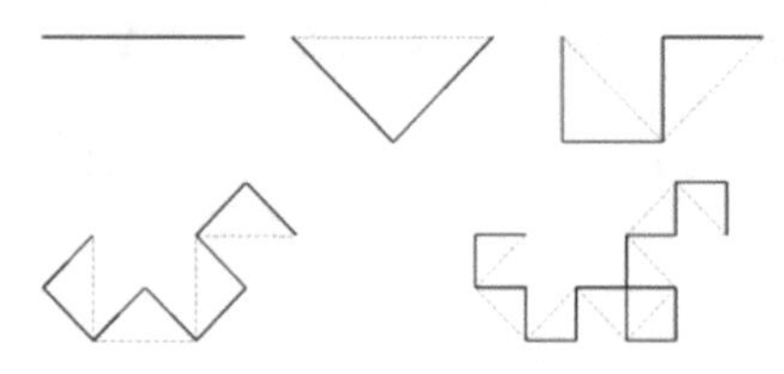

Fig. 5. Harter-Hateway Dragon Construction Algorithm.

However, we don't want to deviate from the original approach of having just one F, interpreted as a step forward. To return to this approach, let's treat X and Y as auxiliary variables that Turtle ignores and replace them with FX and FY, respectively, in the generation rule.

We get: axiom = FX,
FX = FX+YF+,

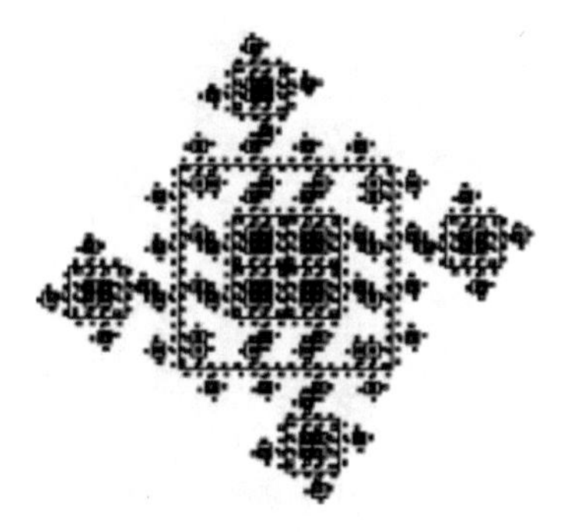

Fig. 6. 3 Repetitive Mosaics (Patrick Hagerty)

YF = -FX-YF.

Additionally, we mention that the same result can be achieved using the following generative rules (Fig. 7):

axiom = FX,

newf = F,

newx = X+YF+,

newy = -FX-Y.

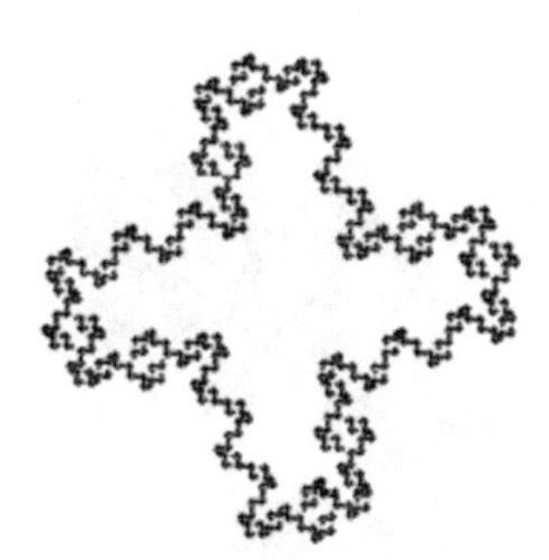

Fig. 7. Chain after 3 reps (Yang-Xi Luo)

The algorithm for creating a fractal image using the L-systems method (Fig. 8):

axiom: "F".

rule: "F" => "FX",

"X" => "X + Y".

"Y" => "FX-Y".

angle: $\frac{\pi}{2}$

Floral pattern fractal.

Axiom: F+F+F+F.

Rule:

F => FF+F+F+F+F+F-F.

Angle: 90°

Fig. 8. Floral pattern fractal

A geometric model and algorithm for generating fractal images using geometric substitutions.

Below some steps are given to build a dragon using these generative rules:

Step 1: FX+YF+

Step 2: FX+YF++-FX-YF+

Step 3: FX+YF++-FX-YF++-FX+YF+ – –FX-YF+

Step 4: FX+YF++-FX-YF++-FX+YF+ – –FX-YF++ -FX+YF++-FX-YF+ – – FX+YF+ – – FX-YF+

Figure 6 shows the Harter-Hateway Dragon after 12 iterations. Note that the dragon consists of several similar parts (Fig. 9).

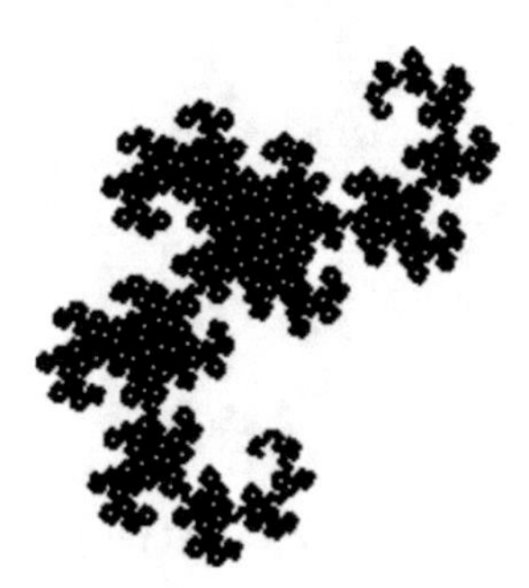

Fig. 9. The result obtained by the method of L-systems.

2. To execute geometric substitutions for the generated fractal image:

$$f_1 = \frac{cos\left(\frac{2\pi}{15}\right)}{cos\left(\frac{(36-n)\pi}{180}\right)} \tag{1}$$

$$f_2 = sin\left(\frac{2\pi}{15}\right) - f_1 \cdot sin\left(\frac{(36-n)\cdot\pi}{180}\right) \tag{2}$$

3. To turn (move) based on a given angle:

$$\{x' = x + rcos\frac{\pi d}{180}y' = x + rsin\frac{\pi d}{180} \text{ where } d = \frac{360^\circ}{n}$$

m – the number of internal reflections, n – the number of external reflections, r – shift (central), itr – the number of steps, d – the turning angle.

3 Results and Discussions

The "Fractal Design" software tool generates fractal images using L-system, IFS, and analytical methods, including geometric permutations.

Figure 10 illustrates the main graphical interface displayed when the software is launched. Here:

1. Name of the software
2. Fractal generation methods
3. Types of forming fractals
4. Fractal generation values
5. Main working area

To build fractals using the L-system method, click the "L-system" button. We need to use the "Tree Fractal 1", "Tree Fractal 2", and "Dragon" buttons for drawing fractals. For instance, when we select the "Dragon" button and click the number of steps, internal reflection, external reflection, and displacement (relative to the center), the result of the L-system method (Fig. 11) will draw a complex fractal image through geometric substitutions.

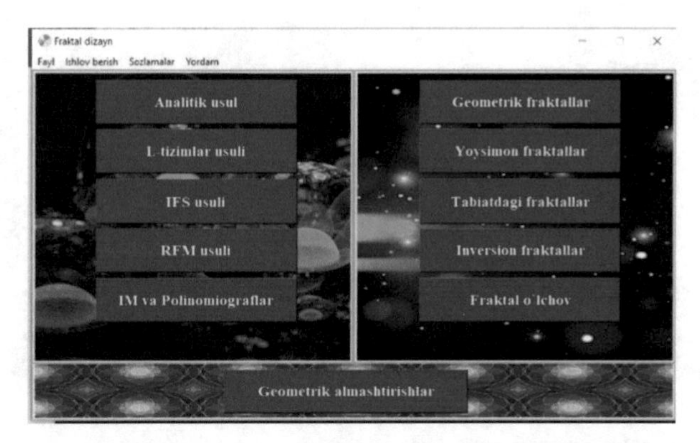

Fig. 10. The main working window of the software

Here, "Int" is the number of steps in the L-systems method, "Number of Reflections" is the number of internal and external displacements and rotations, and "Move" is the distance of the external displacements and rotations to the center. "Int" - 11; "Reflection number" (internal) - 5; "Reflection number" (external) - 8; "Relocation" (relative to the center) - 150; The result is as follows:

Using the "Tree Fractal 1" and "Tree Fractal 2" buttons, we create new complex fractal images by geometrical substitutions of the result obtained by the L-systems method. Click the "Tree fractal 1" button with the following values: "Int" - 4; "Reflection

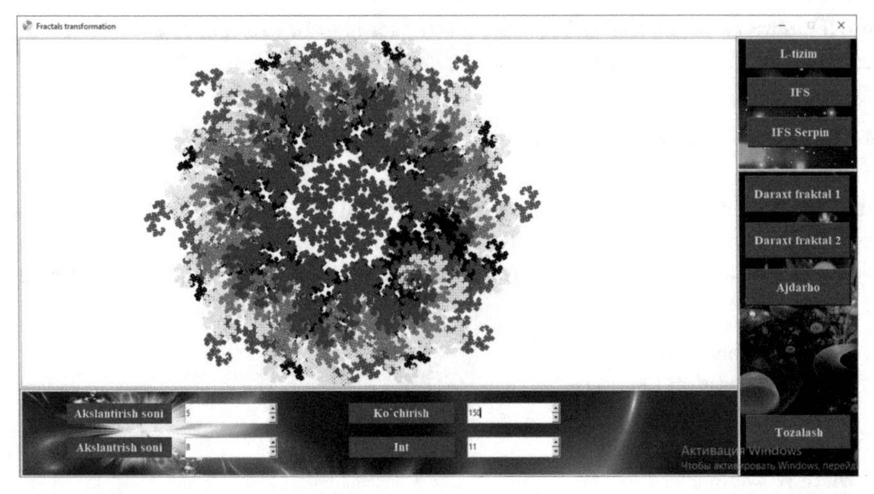

Fig. 11. A new complex fractal representation of the dragon obtained by geometric permutations of the L-systems result.

Fig. 12. A new complex fractal representation obtained by geometrical substitutions of the tree fractal derived from the L-systems method.

number" (internal) - 8; "Reflection number" (external) - 8; "Relocation" (relative to the center) - 250; The result is as follows (Fig. 12):

Click the "Tree Fractal 2" button with the following values: "Int" - 5; "Reflection number" (internal) - 6; "Reflection number" (external) - 6; "Moving" (relative to the center) - 100; The result is as follows (Figs. 13 and 14):

A mathematical model of objects with a fractal structure was built using the L-systems method. Algorithms for drawing and visualization of circular, tree-shaped, star-shaped, and polygonal fractals were developed based on this model. The developed

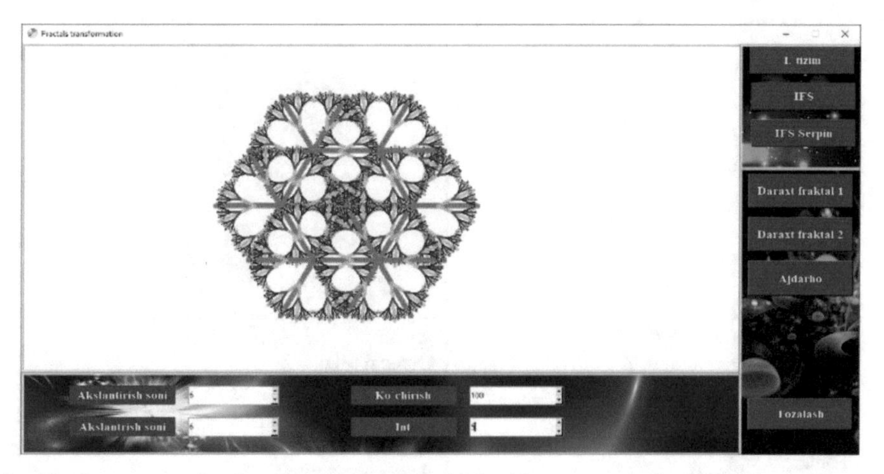

Fig. 13. A new complex fractal representation obtained by geometrical substitutions of the tree fractal derived from the L-systems method.

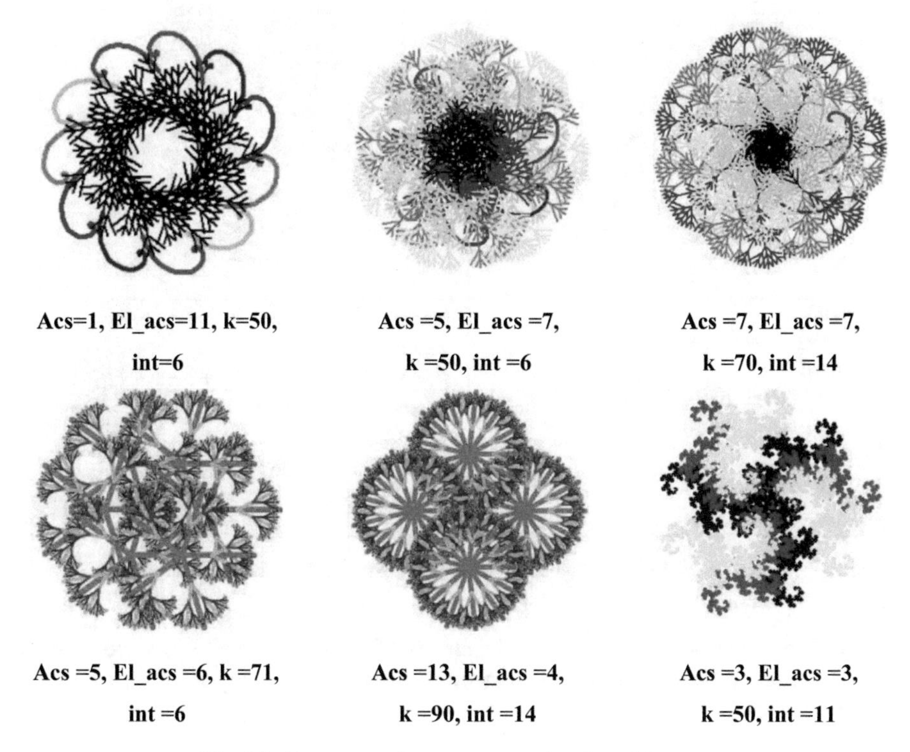

Fig. 14. Dragons and trees are complex fractal images.

algorithms serve to create images with complex fractal structure related to pattern design (Fig. 15).

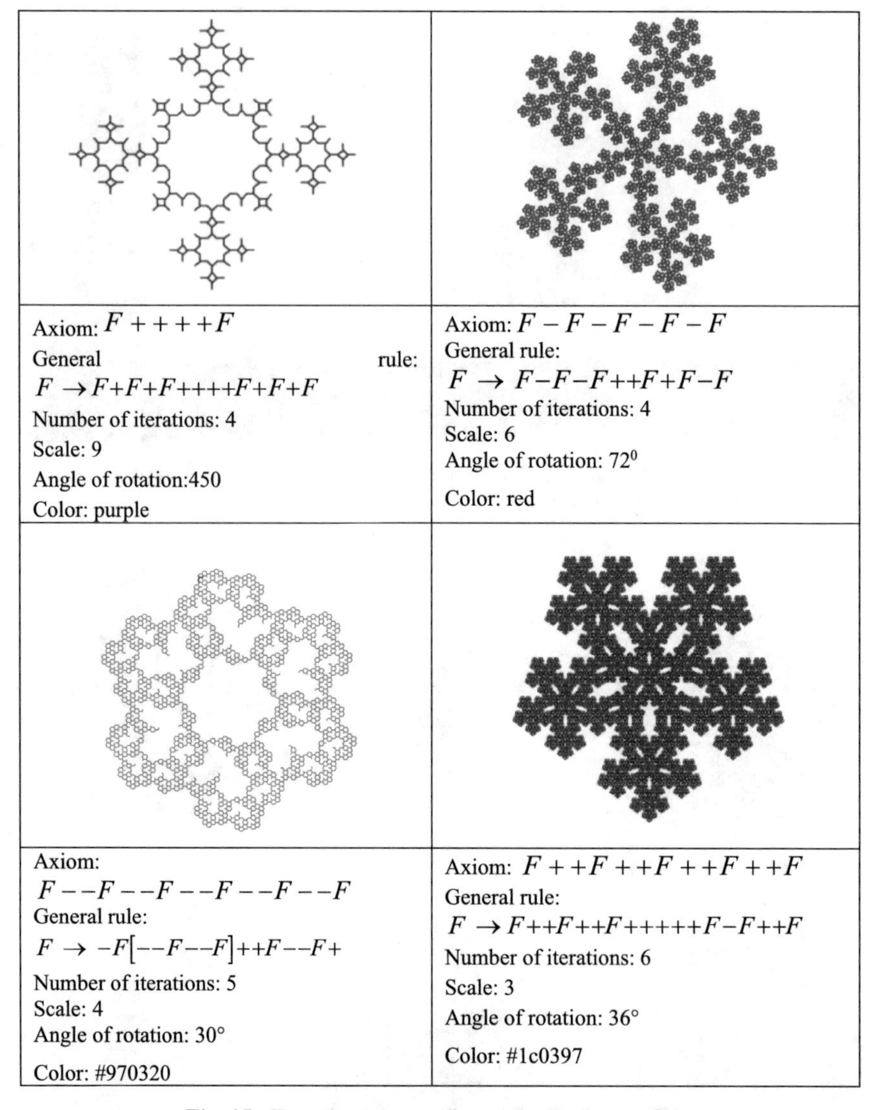

Axiom: $F++++F$ General rule: $F \rightarrow F+F+F++++F+F+F$ Number of iterations: 4 Scale: 9 Angle of rotation:450 Color: purple	Axiom: $F-F-F-F-F$ General rule: $F \rightarrow F-F-F++F+F-F$ Number of iterations: 4 Scale: 6 Angle of rotation: 72^0 Color: red
Axiom: $F--F--F--F--F--F$ General rule: $F \rightarrow -F[--F--F]++F--F+$ Number of iterations: 5 Scale: 4 Angle of rotation: 30° Color: #970320	Axiom: $F++F++F++F++F$ General rule: $F \rightarrow F++F++F+++++F-F++F$ Number of iterations: 6 Scale: 3 Angle of rotation: 36° Color: #1c0397

Fig. 15. Fractals corresponding to the Koch snowflake

The theoretical foundations and advantages of L-systems methods were studied. These methods make it possible to perform a comparative analysis of classical and modern fractals.

4 Conclusion

Fractal shapes are considered to be one of the perfect works created by combining knowledge in mathematics, computer science, and art. Before the introduction of the fractal geometry concept, geometric models of natural objects were described using combinations of simple shapes such as straight lines, triangles, circles, rectangles, and spheres. However, using Euclidean geometry, more complex natural objects are inexplicable, such as porous materials, cloud shapes, and tree shapes. Today, the mathematical aspects of fractal theory, including the methods of describing natural processes and phenomena using the fractal theory, is an independent new field of science. This article defines the software tool developed based on the L-systems method, the algorithm for creating complex fractal structures using geometric substitutions that can be effectively used in the textile industry, in the design of building materials, and in the technologies of printing patterns on gas and carpets.

References

1. Agoston, M.K.: Computer Graphics and Geometric Modeling, vol. 1, pp. 301–304. Springer, London (2005). https://doi.org/10.1007/b138805
2. Hossain, A., Nurujjaman, Akter, R., Ahmed, P.: Fractals generating techniques. Sonargaon Univ. J. **1**(2), 43–54 (2016)
3. Alik, B., Ayyildiz, S.: Fractals and fractal design in architecture. In: 13th International Conference "Standardization, Prototypes and Quality: A Means of Balkan Countries' Collaboration", pp. 3–4 (2021)
4. Anarova, Sh., Narzulloev, O., Ibragimova, Z.: Development of fractal equations of national design patterns based on the method of R-function. Int. J. Innov. Technol. Explor. Eng. **9**(4), 134–141 (2020)
5. Ben, T., Cecilia, B., Brian, B.: Math Coloring Book Fractals, pp. 60–68 (2018)
6. Bourke, P.: An introduction to the Apollonian fractal. Comput. Graph. **30**(1), 134–136 (2006)
7. Branch, M., et al.: Fractal geometry and persian carpet. In: Bridges 2012: Mathematics, Music, Art, Architecture, Culture, pp. 457–460 (2012)
8. Barnsley, M.F.: Fractals Everywhere. Academic Press, New York (2014)
9. Chugh, R., Kumar, V., Kumar, S.: Strong convergence of a new three step iterative scheme in Banach spaces. Am. J. Comput. Math. **2**(04), 345 (2012)
10. Clancy, C., Frame, M.: Fractal geometry of restricted sets of circle inversions. Fractals **3**(04), 689–699 (1995)
11. Edgar, G.: Measure, Topology, and Fractal Geometry, pp. 18–25. Springer, Heidelberg (2007). https://doi.org/10.1007/978-0-387-74749-1
12. Falconer, K.: Fractal Geometry: Mathematical Foundations and Applications, pp. 180–183. Wiley, Hoboken (2004)

Predicting the Popularity of Social Network Publications Based on Content Analysis Using the Transformer Language Model

Maksim Shishaev(✉) and Vladimir Dikovitsky

Institute for Informatics and Mathematical Modeling, Kola Science Centre of the Russian Academy of Sciences, Apatity, Russia
shishaev@iimm.ru

Abstract. The paper examines the possibility of using pre-trained neural network language models of the transformer architecture to predict the popularity of messages in online social networks. The authors used a pre-trained GPT-2 network to solve the problem of classifying messages by popularity based on a dataset formed from messages from several virtual communities of the VKontakte social network. The resulting classifier demonstrated an accuracy of over 70%. The number of likes normalized by the number of views was used as a popularity metric.

Keywords: Social Network · Popularity Prediction · Transformer Language Model

1 Introduction

Online social networks are one of the most popular and influential communication channels in the modern world. They allow users to create and distribute various types of content such as texts, images, videos, etc. Miscellaneous content analysis of online social networks is used to solve many current applied problems [1]. Content that is published on social networks can have varying degrees of popularity, which depends on many factors, such as topic, quality, time of publication, audience, etc. The popularity of content is of great importance for both users and administrators of social networks. For users, the popularity of content can be a motivation for creating new content, an indicator of audience interest and involvement, and a means of increasing their status and reputation in a social network. For social network administrators, content popularity can be a criterion for ranking and recommending content, a tool for analyzing user behavior and preferences. To solve such applied problems, it is necessary to be able to predict the popularity of publications based on some observable features, which can be both the inner properties of the publication's content and external features that characterize the context of the publication's appearance and distribution.

This paper explores the possibility of predicting the popularity of a publication on an online social network solely based on its content. The approach under consideration is based on the assumption that modern pre-trained language models based on the

A. Gibadullin (Ed.): DITEM 2023, LNNS 942, pp. 180–191, 2024.
https://doi.org/10.1007/978-3-031-55349-3_15

transformer architecture provide some internal formal representation of the meaning and other properties of the content, which can be used as a classifying feature of the popularity of a publication. As such a language model, the work used GPT-2, which was additionally trained to solve the problem of classifying text messages by popularity based on the labeled data set available to the authors, collected from the social network VKontakte. This assumption is based on the fact that the GPT models demonstrate high efficiency in various natural language processing tasks, such as question answering, machine translation, reading comprehension, and summarization [2].

2 Materials and Methods

2.1 A Brief Overview of Approaches to Analyzing the Popularity of Content on Social Networks

Predicting the popularity of content on online social networks is a relevant and challenging task that attracts the attention of many researchers. Existing works on this topic can be classified according to the type of data used, analysis methods, and research objectives. One of the main criteria for classifying works on content popularity analysis is the type of data used. Depending on this, the following approaches can be distinguished:

Content Analysis. This approach focuses on analyzing the content of messages distributed on social networks, while the message can have various formats - text, image, video, etc. The goal of this approach is to identify the properties and characteristics of content that influence its popularity. For example, in [3], the authors analyze textual features of news articles, such as title, length, complexity, sentiment, etc., and their relationship with the number of comments on Facebook.

User Analysis. This approach is based on the analysis of the properties of users involved in the distribution of content on social networks, such as authors, recipients, intermediaries, etc. The goal of this approach is to identify the properties and characteristics of users that influence the popularity of content. For example, in the work [4] the authors analyze the characteristics of Twitter users, such as the number of subscribers, frequency of tweets, level of activity, etc., as well as their relationship with the number of retweets and likes. In the work [5] the authors study the characteristics of YouTube users, such as gender, age, geography, etc., and their impact on the number of views and comments on videos.

Formal Analysis of a Social Network. This approach is based on SNA (Social Network Analysis) methods and focuses on analyzing the structure and dynamics of the properties of the network of social contacts in which content is distributed - such as topology, density, clustering, etc. The goal of this approach is to identify network properties and characteristics that influence content popularity. For example, in the work [6] the authors analyze the structure of the network of reposts of news articles on Facebook and its relationship with the number of comments and likes. In [7] the authors study the dynamics of the video distribution network on YouTube and its impact on the number of views and comments.

It can be noted that in the considered existing approaches to predicting content popularity, both heterogeneous objects (articles, users, networks of interactions, etc.)

and multiple heterogeneous features are used. This makes them difficult to use, since it requires significant expenses for preparing and marking the corresponding datasets. In contrast, in our work we use only the text of messages as such, without any additional features.

2.2 Formal Representation of Popularity

To formally represent the popularity of a message, it is necessary, first of all, to determine the conceptual understanding or applied interpretation of this property. The interpretation of this term, in turn, relies on the conceptual representation of the problem. A fragment of such a representation, which includes the most significant components of the subject area in the context of this work, is presented in the form of a conceptual graph in Fig. 1. In this article, we proceed from the conceptual interpretation of popularity as a measure of the intensity of the audience's reaction to a message (publication).

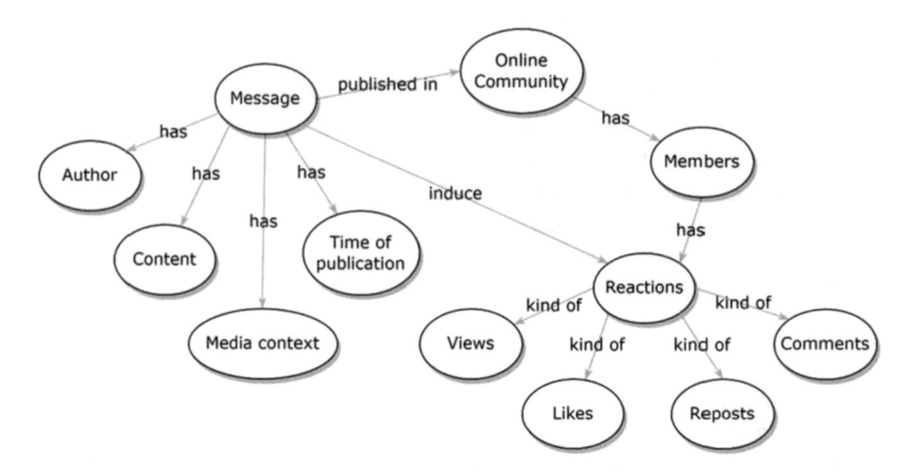

Fig. 1. Conceptual graph of the subject area.

The formal representation of popularity is based on some observable and measurable features of publication. These are usually the number of views, the number of reposts, the number of comments and the number of likes. That is,

$$Popularity = f(s, r, c, l) \tag{1}$$

where s, r, c, l – the number of views, reposts, comments and likes, respectively.

The form of the function f depends on the interpretation used of the concept of popularity of a publication, which, in turn, is determined by the applied problem in the context of which we need to quantify popularity. As a result, there is no common understanding of popularity in existing works on this topic, which makes its formal definition a non-trivial task [8]. In addition to the fact that the popularity of a publication can be expressed in different ways, it can change over time, depending on various unformalized factors [9]. Another factor that complicates the task is that the popularity of identical content may vary across different social networks or even within different groups within the same network. Authors of [3] to predict popularity, resort to using other parameters of publications, for example, homophily of the audience and the popularity of the author of publications. Another approach to determining the popularity of content is based on analyzing its distribution on social networks. The work [10] proposes to exploit differences in audience behavior and interests. The authors use data from Twitter and YouTube and show that group models outperform general models in predicting content popularity accuracy. The authors demonstrate that their method is able to predict the impact of content on users with high accuracy and interpretability.

In addition to quantitative indicators of the audience's reaction to a publication, various modifiers can be used when calculating the score, for example, the astronomical time of publication of the message or number of subscribers of the online community in which the message was published. An important consideration is also whether the popularity of messages is assessed within one community or several. In the latter case, obviously, estimates in absolute terms are not suitable, since the quantitative values of popularity indicators, other things being equal, directly depend on the size of the audience of the online community.

Since in this work we focused on tasks related to monitoring several online communities at once, relative metrics were used to assess popularity, representing the ratio of the values of audience reaction indicators to the number of views. The only open question is which indicators (likes, reposts or comments) or combinations thereof to use.

In this work, popularity prediction was interpreted as a classification problem, which implies two main stages - (1) determining the value of the object's classifying feature (in our case, the message's popularity metric) and (2) assigning the object to one of two classes (popular/not popular) in accordance with whether the value of the attribute overcomes a given threshold. With this formulation of the problem, it is obvious that it is not the value of the attribute as such that is decisive, but the order of the objects sorted by this attribute. In this sense, methods for calculating the values of object attributes that result in an identical sorting order for objects are equivalent. To test this statement, we used formerly accumulated experimental data. The objects (messages) were sorted by number of likes, reposts and comments and the degree of similarity of the resulting sequences was assessed.

Sequence similarity was assessed as the average relative displacement of the object position in two sequences:

$$s = \left(\sum\nolimits_{i=1}^{N} \frac{p_1(x_i) - p_2(x_i)}{N} \right) / N \tag{2}$$

where N is the number of objects, $p_1(x_i), p_2(x_i)$ is the position of object x_i in the 1st and 2nd sequences, respectively.

The size of the initial dataset was 100,000 samples. During the comparison, only samples with non-zero values of the corresponding labels were considered.

Six different sequences were formed: P_{lc}, P_{cl}, P_{lr}, P_{rl}, P_{cr}, P_{rc}. The subscript in the sequence designation in this case means the following: P_{lc}, - a sequence sorted in descending order l, in which samples have a non-zero number of likes and comments.

The designations of the remaining sequences are interpreted in a similar way. That is, the following sets of conditions are met:

$$\begin{cases} o \in P_{lc} : likes(o) \neq 0 \wedge comments(o) \neq 0 \\ \forall o_i, o_j \in P_{lc} : likes(o_i) \geq likes(o_j), \forall i < j \end{cases} \quad (3)$$

$$\begin{cases} o \in P_{cl} : likes(o) \neq 0 \wedge comments(o) \neq 0 \\ \forall o_i, o_j \in P_{cl} : comments(o_i) \geq comments(o_j), \forall i < j \end{cases} \quad (4)$$

$$\begin{cases} o \in P_{lr} : likes(o) \neq 0 \wedge reposts(o) \neq 0 \\ \forall o_i, o_j \in P_{lr} : likes(o_i) \geq likes(o_j), \forall i < j \end{cases} \quad (5)$$

$$\begin{cases} o \in P_{rl} : likes(o) \neq 0 \wedge reposts(o) \neq 0 \\ \forall o_i, o_j \in P_{rl} : reposts(o_i) \geq reposts(o_j), \forall i < j \end{cases} \quad (6)$$

$$\begin{cases} o \in P_{cr} : comments(o) \neq 0 \wedge reposts(o) \neq 0 \\ \forall o_i, o_j \in P_{cr} : comments(o_i) \geq comments(o_j), \forall i < j \end{cases} \quad (7)$$

$$\begin{cases} o \in P_{rc} : reposts(o) \neq 0 \wedge comments(o) \neq 0 \\ \forall o_i, o_j \in P_{rc} : reposts(o_i) \geq reposts(o_j), \forall i < j \end{cases} \quad (8)$$

where *likes(o), reposts(o), comments(o)* are the number of likes, reposts and comments of sample o, respectively, o_i is the *i*-th element of the sequence.

Sequences were compared in pairs. The size of the corresponding sequences and the results of the comparison are shown in Table 1.

Table 1. Comparison of sequences obtained by element selection and sorting.

Sequences compared	Sequence size	Average relative displacement of an object's position over the entire sequence
P_{lc}, P_{cl}	66970	0.278
P_{lr}, P_{rl}	80310	0.191
P_{cr}, P_{rc}	59397	0.248

It can be noted that the difference between the sequences of samples sorted by likes and reposts does not exceed 20% (in terms of the relative displacement of elements). At the same time, sequences formed by sorting by the number of comments differ significantly from others - by 25% or more. Thus, using the number of likes or the number of reposts as a basis for calculating a message's popularity score will potentially produce an equivalent result.

It can also be assumed that the observed features of the sequences are due to significant differences in the quantitative values of indicators of audience reaction - likes, comments and reposts. For the dataset used in the work, the average values of these indicators were 53.7, 9.7 and 13.8, respectively. This assumption is also confirmed by the type of dependence of the relative displacement of objects in sequences on their response indicator values. These dependencies for the sequences P_{lc}, P_{cl}, P_{lr}, P_{rl}, P_{cr}, P_{rc} are presented in Fig. 2. To construct them, a series of subsequences were formed in each sequence, including K initial elements of the original sequence - $\{o_1, \ldots, o_K\}$, and for each subsequence the average relative displacement of the elements was calculated.

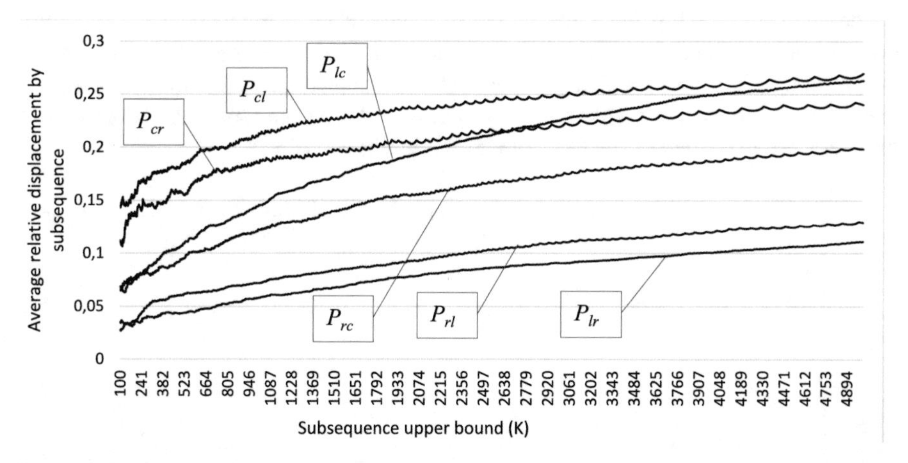

Fig. 2. Dependence of the average relative displacement of objects in sequences on the values of response indicators (cumulative total).

Since the sequences are sorted in descending order of the corresponding reaction indicators, the sequences that include elements with the highest values of likes, reposts, and comments are concentrated at the beginning of the x-axis. It can be seen that for sequences that include objects with large indicator values, the average relative displacement is lower. Thus, for posts with high reaction indicators, using either of them as the basis for calculating the popularity score will produce a similar result.

Since the absolute values of the "number of likes" indicator are higher than other indicators, this work used this indicator, additionally normalized by the number of content views:

$$popularity(likes, views) = \begin{cases} 0, 0 < est < k \\ 1, est \geq k \end{cases}, where\ est = \frac{likes}{views} * 100 \quad (9)$$

The parameter k in formula (9) means the threshold of classification feature. To determine its optimal value, a series of experiments was carried out, described in the next section.

3 Results

3.1 Dataset Description

For the experiment, previously collected data on publications (posts) in several open regional online communities of the social network "VKontakte" for the period from January 2021 to December 2022 were used. A single post (message) was considered as a data sample, and its attributes were the number of likes, comments and reposts. The collection included communities without a specific thematic connection, but localized by the constitution of subscribers in a small region (in our case, the Kirovsk-Apatity subregion of the Murmansk region). Thus, the data set used for the experiments can be characterized by a variety of topics of publications, and a relatively stable composition of the audience, mainly from among the residents of the region under consideration. For a meaningful analysis of the dataset used in the work, its thematic modeling was carried out, as a result of which the most discussed topics were identified. The sizes of the obtained topics (the number of messages included in the cluster corresponding to the topic) are presented in the form of a graph in Fig. 3.

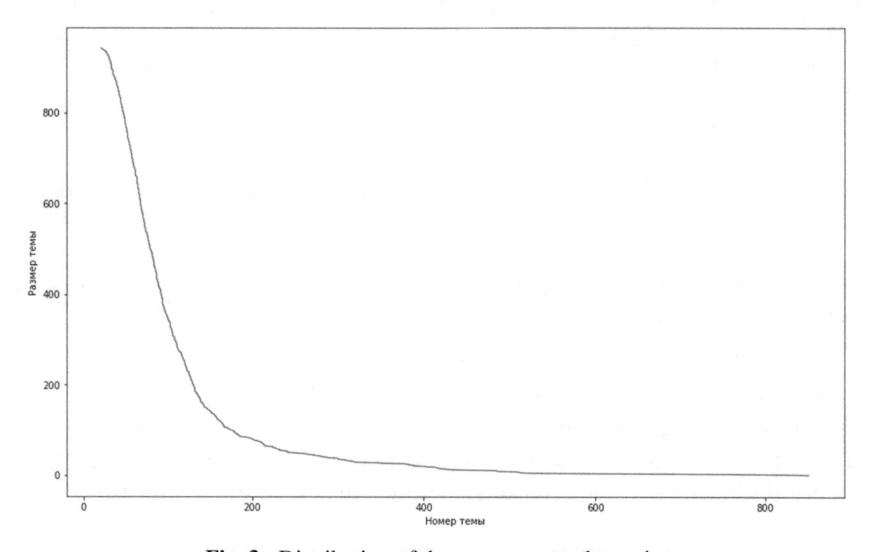

Fig. 3. Distribution of dataset messages by topic.

In total, the dataset included data from 20 online communities with 296,000 active subscribers, which ensures the representativeness of the data in the context of the intended practical use of the results obtained. In addition, to test the resulting classifier on external data, a dataset with similar parameters was used, collected in communities of another region of similar scale.

A peculiarity of online social networks is the dominance of short messages (the distribution of posts by length is shown in Fig. 4). Since the reaction to a publication was predicted only by the content of the message, reflecting its meaning, to maintain the purity of the experiment, posts with a typical length were selected from the entire sample, in this case up to 100 words.

Data preprocessing included the following steps:

- Removing short posts that contain no text or only contain links or hashtags, as they are not suitable for analysis using language models.
- Removing long posts that contain more than 100 words because they are anomalous or irrelevant to the current study.

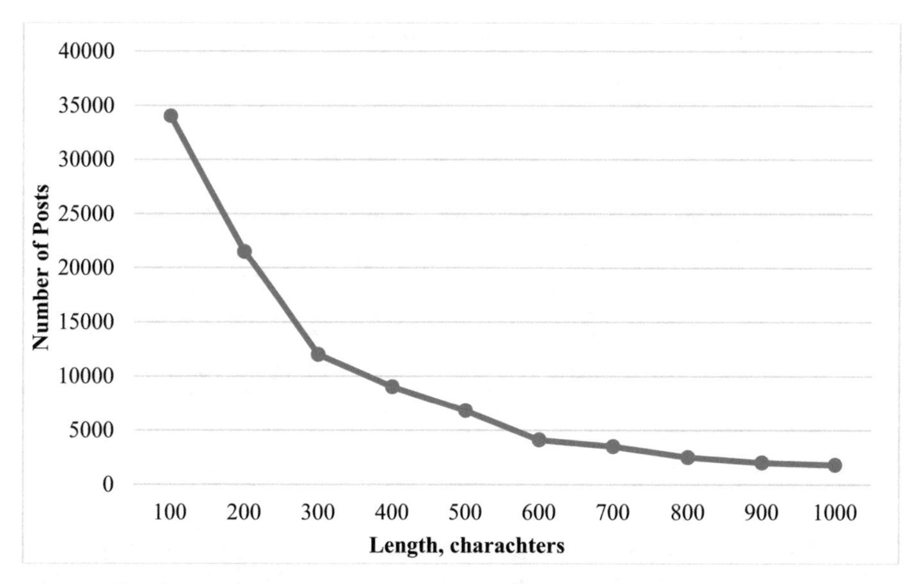

Fig. 4. Number of samples in the collection depending on message length.

- Removing posts that contain profanity, insults, spam, as they can distort the results of the analysis.
- Bringing the text of posts to lower case, removing punctuation marks, emoji and other special characters that do not affect the meaning of the text.

3.2 General Scheme and Results of the Experiment

To classify popular and unpopular messages, it is proposed to use a trainable classifier that is capable of identifying various weakly formalized criteria from the text of posts, such as style, tone, relevance, virality, originality, etc. The generative language model GPT was chosen as the base model for the classifier, which shows good results in text analysis (summarization) tasks [2]. To analyze the text of social network publications in order to classify them into popular and unpopular, an autoregressive generative language model based on transformer architecture (GPT-2) was used, which is capable of taking into account the context and semantics of the text, as well as generating new texts based on trained parameters.

The GPT-2 model was fine-tuned on a corpus of labeled additional data containing information about the number of likes, comments and reposts for each message. Thus, the GPT-2 model was adapted to the specifics of the task of identifying the popularity of messages in social networks. As the base, the GPT-2 model was used, pretrained on a data set of 8 million web pages and containing 1.5 billion parameters.

As stated earlier, in this work, the problem of popularity prediction is considered as a two-class classification problem. In this case, the value of the classification characteristic of samples depends on the value of the threshold value k (see formula 9). A series of experiments were conducted to train the classifier at different threshold values. For this purpose, 5 different training sets were generated, the characteristics of which are given in Table 2. The sets were balanced by under-sampling - removing excess instances of the dominant class (in our case, negative).

Table 2. Characteristics of datasets.

Threshold	Positive samples count	Negative samples count	Training subset		Test subset	
			Positive	Negative	Positive	Negative
1	20675	79149	15507	15508	5168	5169
2	8332	91492	6250	6251	2082	2083
3	3250	96574	3654	3718	998	929
4	1086	98738	815	816	271	272
5	359	99465	270	271	89	90

The classifier training results for different threshold values are presented in Table 3.

The results obtained show that the optimal threshold value is 3 (with a threshold value of 5, the classification accuracy is higher, but the size of the dataset is significantly reduced).

For testing on external data that was not used during training, the resulting classifier was used to predict the popularity of messages from an alternative dataset collected from virtual communities in another region (Volkhov, Leningrad region). The test dataset contained 1062 and 1061 instances of the positive and negative classes, respectively. The test results are presented in Table 4.

The deterioration of the result when testing on external data can be explained by the difference in the thematic structure of public discourse in one and another region.

Table 3. Classifier training results.

Threshold value	Metric/Class	Precision	Recall	F1 score
1	Negative	0.65	0.78	0.71
	Positive	0.72	0.57	0.64
	Accuracy			0.68
	Weighted average	0.69	0.68	0.67
2	Negative	0.66	0.83	0.73
	Positive	0.76	0.57	0.65

(continued)

Table 3. (*continued*)

Threshold value	Metric/Class	Precision	Recall	F1 score
	Accuracy			0.70
	Weighted average	0.71	0.70	0.69
3	Negative	0.82	0.60	0.69
	Positive	0.70	0.88	0.78
	Accuracy			0.74
	Weighted average	0.76	0.74	0.73
4	Negative	0.62	0.70	0.66
	Positive	0.65	0.56	0.60
	Accuracy			0.63
	Weighted average	0.63	0.63	0.63
5	Negative	0.72	0.82	0.77
	Positive	0.79	0.67	0.73
	Accuracy			0.75
	Weighted average	0.75	0.75	0.75

Table 4. Results of testing on the external dataset.

Threshold value	Metric/Class	Precision	Recall	F1 score
3	Negative	0.50	0.96	0.66
	Positive	0.58	0.06	0.1
	Accuracy			0.51
	Weighted average	0.54	0.51	0.38

4 Discussion

Based on the results obtained, we can conclude that the language model based on the transformer architecture makes it possible to predict the popularity of content with good accuracy. Using data from a limited virtual community, this accuracy exceeds 70%, and on external data it reaches 51%. Thus, within the community whose data was used for training, the model can be considered suitable for practical use. The advantage of the proposed method, in comparison with other approaches to predicting popularity, is the use of only the content of messages as such, which eliminates the need to implement complex procedures for generating labeled data sets and training the model.

At the same time, it seems possible to increase the accuracy and expand the scope of popularity prediction based on language models of the transformer architecture. In this regard, the following assumptions can be formulated:

- The intensity of the reaction depends on the daytime at which the message is published. At the same time, for different online communities and categories of users, the "prime time" that provides the most intense reaction in absolute terms will be different - in some cases it will be evening hours, in others it will be during the day or morning or even at night.
- The intensity of the reaction to a message depends on the information and media context of the latter's appearance. This context characterizes the range of the most significant topics in public discourse at the time the message appears. For example, at the beginning of the Covid-19 pandemic, any message on this topic was much more popular than now, when the severity of the coronavirus problem has subsided.
- A potential increase in predicting accuracy is possible by taking into account the nature of the message topic when training and using the model - is this a viral topic causing bursting reaction, or so-called "greenline" topic with stable in time popularity.
- The ability of the GPT model predicting the popularity of a message only based on the content of the latter allows us to consider such models as a component of generative adversarial networks focused on generating content that causes the expected reaction of the audience.

5 Conclusion

This paper examines the possibility of using a pre-trained neural network language model of the Transformer architecture to predict the popularity of messages in online social networks. The experiments conducted, based on the dataset available to the authors, formed from messages from several localized online communities of the VKontakte network, showed that the autoregressive model of the transformer architecture is capable of quite accurately determining and, therefore, predicting the popularity of content in social networks.

During the work, special attention was paid to the issue of formalizing the popularity indicator. An auxiliary study conducted showed that the audience's reaction is approximately equally exhibited in both the number of likes and the number of reposts and comments on the message. However, when predicting the popularity of messages from different online communities, it is necessary to use relative metrics based on the number of subscribers or the number of views.

A promising direction for continuing work seems to be the analysis of the influence of additional factors on the popularity of content, such as topic, format, virality, publication time, etc. An urgent task is also to identify the characteristics and differences in the popularity of content in different audience subgroups.

References

1. Shishaev, M., Fedorov, A., Datyev, I.: Analysis of online social networking when studying the identities of local communities. In: Salminen, M., Zojer, G., Hossain, K. (eds.) Digitalisation and Human Security. NSC, pp. 267–293. Springer, Cham (2020). https://doi.org/10.1007/978-3-030-48070-7_10
2. Radford, A., Wu, J., Child, R., Luan, D., Amodei, D., Sutskever, I.: Language models are unsupervised multitask learners. OpenAI blog (2019)

3. Shang, Y., Zhou, B., Zeng, X., Wang, Y., Yu, H., Zhang, Z.: Predicting the popularity of online content by modeling the social influence and homophily features. Front. Phys. **10** (2022)
4. Pastuhov, R.K., Drobyshevskij, M.D., Turdakov, D.: Identification of influential social network users using a bipartite comment graph. Proc. Inst. Syst. Program. RAN **34**, 127–142 (2022)
5. Brodovskaya, E., Dombrovskaya, A.: Big data in political process research. Litres (2022)
6. Bakshy, E., Messing, S., Adamic, L.A.: Exposure to ideologically diverse news and opinion on Facebook. Science **348**, 1130–1132 (2015). https://doi.org/10.1126/science.aaa1160
7. Hoiles, W., Aprem, A., Krishnamurthy, V.: Engagement and popularity dynamics of youtube videos and sensitivity to meta-data. IEEE Trans. Knowl. Data Eng. **29**, 1426–1437 (2017). https://doi.org/10.1109/TKDE.2017.2682858
8. Tatar, A., de Amorim, M.D., Fdida, S., Antoniadis, P.: A survey on predicting the popularity of web content. J. Internet Serv. Appl. **5**, 8 (2014). https://doi.org/10.1186/s13174-014-0008-y
9. Szabo, G., Huberman, B.A.: Predicting the popularity of online content. Commun. ACM **53**, 80–88 (2010). https://doi.org/10.1145/1787234.1787254
10. Ding, K., Wang, R., Wang, S.: Social media popularity prediction: a multiple feature fusion approach with deep neural networks. In: Proceedings of the 27th ACM International Conference on Multimedia, pp. 2682–2686. Association for Computing Machinery, New York (2019). https://doi.org/10.1145/3343031.3356062

Classification of Building Components Using Artificial Intelligence Based on Meta-Information in BIM Models

M. Motina(✉) and A. Morozenko

Moscow State University of Civil Engineering, 26 Yaroslavskoe shosse, Moscow 129337, Russia
mvmotina@yandex.ru

Abstract. Today there is an urgent need for digitalization of many processes in the construction industry, but the problems of automation of some construction processes have not yet been implemented in many companies due to lack of resources. This article discusses the methodology of classification of building elements using artificial intelligence (AI) and analysis of meta-information within each element. The article presents a detailed analysis of the use of TF-IDF and logistic regression for the classification of building components, as well as an assessment of the effectiveness of this technique along with traditional methods. This method can be used to automate construction processes throughout the entire life-cycle of the construction of an object in the presence of metadata of classification objects. The study also covers the importance of this method for optimizing processes in the construction industry and improving the quality of projects.

Keywords: Artificial Intelligence · TF-IDF · Logistic regression · Classification of Building Components · Meta-Information · Life-Cycle of the Construction · BIM · Construction · Machine Learning

1 Introduction

In this article, it is proposed to consider the experience of creating a machine learning model for analyzing big data from digital asset models at the design stage to solve the problem of automation of construction planning (see Fig. 1).

It is expected that artificial intelligence will significantly change the processes of creating added value in the construction and real estate industry. First of all, it will affect processes in which it is required to analyze large amounts of data, identify possible risks and deviations. These changes will be noticeable at the stages of planning, design, interaction with suppliers, construction management, operation and maintenance of facilities.

The greatest effect from the introduction of AI is expected at the stages before construction: design, analysis and modeling, as well as preparation for construction. AI is designed to solve complex intellectual tasks based on database analysis. Machine learning allows you to identify and eliminate potential problems faster than a human, which makes it an indispensable tool in this industry.

A. Gibadullin (Ed.): DITEM 2023, LNNS 942, pp. 192–205, 2024.
https://doi.org/10.1007/978-3-031-55349-3_16

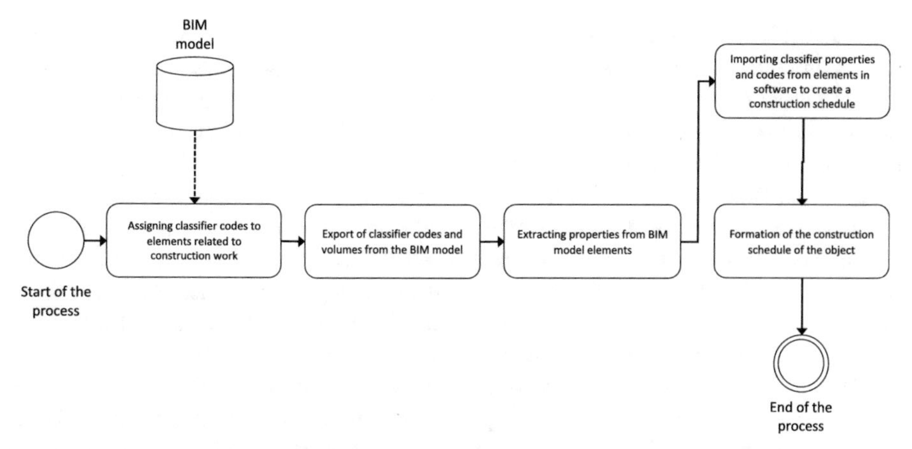

Fig. 1. A scheme of the process of creating a construction schedule for an object based on data from building information models using the classifier code.

2 Literature Review

According to the results of the research conducted by Deloitte last year, more than 75% of the heads of construction and design organizations representing various countries of the world expressed their intention to invest in a variety of digital technologies this year [1].

Researchers A.V. Gusev and S.L. Dobridnyuk noted that with the use of artificial intelligence and digital technologies, a unique ability to make decisions and interact with a person is achieved, providing management with high-level professional tools for observation, design, forecasting, management and control [2].

The research by Solihin W., Eastman C. states that the use of the construction information classifier will form the basis for automated verification of digital models, which as a result will allow us to switch to the format of project documentation that allows us to work with machine analysis and data processing [3].

Petrochenko M.'s research also talks about the successful experience of learning neural models on world and Russian classifiers using various machine learning models for classification tasks: RF-Random Forest, LightGBM, XGBoost, CatBoost [4].

3 The Use of Artificial Intelligence in the Analysis of Meta-Information

3.1 Materials and Methods

The classification process in project activities can be organized by the following method.

An automatic method using artificial intelligence algorithms. With the development of machine learning and artificial intelligence, more effective classification methods have appeared. For example, machine learning algorithms such as the Support Vector Machine (SVM) method, RF-Random Forest [5], LightGBM [6], XGBoost [7, 8],

CatBoost [9]. Also, text analysis methods such as TF-IDF (Term Frequency-Inverse Document Frequency) are used to extract features from text data, which can include the task of classifying BIM-elements. It allows you to determine the importance of terms in textual descriptions of elements. Additionally, logistic regression is used to create AI models capable of classifying elements based on calculated features. The application of the proposed methodology in the construction industry can lead to a significant improvement in the efficiency and accuracy of the classification of building elements. This can be useful in planning and managing construction projects, as well as in architectural design. Similar AI models have already been implemented in medicine (assistance in making a diagnosis by describing symptoms [10]), in the classification of scientific articles [11], determining the mood level in comments on websites [12], and so on.

This article will consider the use of artificial intelligence using TF-IDF and logistic regression methods to classify building elements based on meta-information. These methods can significantly improve the efficiency and accuracy of classification in the construction industry, contributing to the automation and optimization of processes.

4 Application of TF-IDF and Logistic Regression in the Classification of Building Elements

The goal of the study was to evaluate the possibility of using machine learning models based on TF-IDF and logistic regression methods for classifying construction information.

Research tasks:

- Processing BIM model data to create a learning sample for a machine learning model;
- Determination of the minimum achievable accuracy when the AI model is working on the object on which it was learned;
- Determination of the minimum achievable accuracy and the number of iterations of additional learning when the AI model is working on a completely new object;
- Create a list of recommendations for creating similar AI models.

During the research, the following criteria for evaluating the results were identified:

Precision. In machine learning, it is a metric used to evaluate the performance of an AI classification model. It measures the proportion of correct predictions made by the AI model relative to the total number of examples in the test dataset (1):

$$Precision = \frac{The\ number\ of\ correct\ predictions}{Total\ number\ of\ examples} \tag{1}$$

F1-Score (Weighted). A metric used to assess the quality of classification AI models, especially when the classes in the data are unbalanced classes. It takes into account both the accuracy and recall for each class, and then calculates the average value, taking into account the share of each class in the total data set (2):

$$F1_{weighted} = 2 \times \frac{Precision + Recall}{Precision \times Recall} \tag{2}$$

where *Precision* - shows the proportion of correctly classified objects among all objects that the classifier has assigned to this class, *Recall* - displays the proportion of correctly classified class objects to the total number of elements of this class.

F1-Score (Macro). A metric used to evaluate the quality of AI classification models. F1-macro is calculated by averaging F1-score for each class without taking into account their share in the data (3):

$$F1_{macro} = \frac{(F1_1 + F1_2 + \ldots + F1_n)}{n} \tag{3}$$

where $F1_i$ - the F1-score for the i-th class, n - the total number of classes.

Problem statement: it is necessary to create and learn an AI model to automatically fill in the classifier of building elements. The facility "Hotel" was chosen for learning.

Facility characteristics:

- Facility type: residential building;
- Total area of the building: 51 721 sq. m;
- Number of floors: 4;
- Number of rooms: 304 units;
- Number of BIM-models: 45 units;
- Total number of elements: 532 399 units;
- Parts: structural sections, architectural sections and engineering networks sections.

The classifier used to learn the AI model consists of 72 positions (see Fig. 2). For ideal conditions, since the classifier capacity is small, it is enough to learn the AI model on one facility, within which each position of the classifier will occur at least 300 times.

However, the classifier was created for different types of facilities, so some positions are rare in this type of facility, which should be borne in mind.

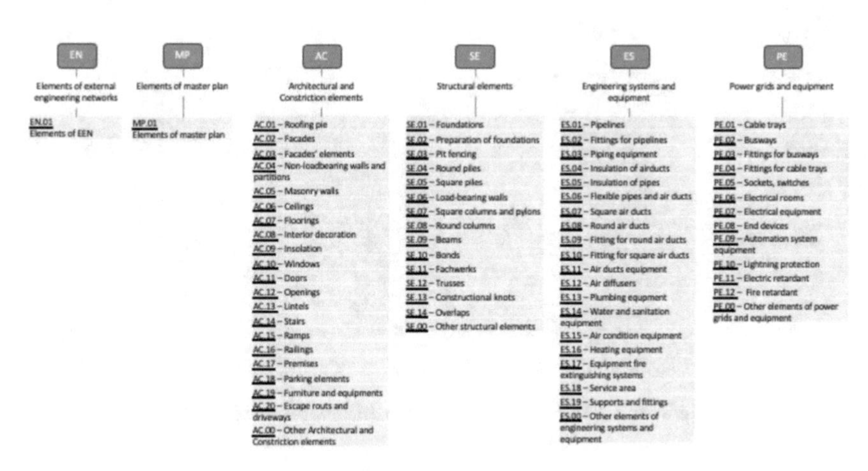

Fig. 2. Classifier Structure.

The facility "Company's Office" was selected for testing the AI model.

Facility characteristics:

- Facility type: office building;
- Total area of the building: 120 800 sq. m;
- Number of floors: 22;
- Number of BIM-models: 67 units;
- Total number of elements: 845 698 units;
- Parts: structural sections, architectural sections and engineering networks sections.

4.1 Data Preparation and Meta-Information Extraction

Before starting the development of an AI model based on TF-IDF and logistic regression, it is necessary:

Mark up 45 Units of BIM Models of the Learning Facility "Hotel". The markup involves filling in classification data in each element (532 399 units) (see Fig. 4). To complete this stage, BIM models of each section created by the project organization were taken, and Autodesk Revit 2021 software was also used.

It was determined by expert means that to fill in the attribute "Assembly code" in each element, it is enough to create specifications for each category of elements that are present in the BIM model. The specifications were created based on the information inside each element recorded in the following attributes: Id, Category, Family, Type (see Fig. 3). This information is sufficient to determine which class of the 72 positions of the classifier the element belongs to.

<Specification for several categories>

A	B	C	D	E
Category	Family	Type	Assembly code	Description of Assembly code
Air Ducts equipment	(DA)_SingleLeafDamper_Square_ABK_Arctic	ABK_150x100	ES.11	Air Ducts equipment
Air Ducts equipment	(DA)_SingleLeafDamper_Square_ABK_Arctic	ABK_200x150	ES.11	Air Ducts equipment
Air Ducts equipment	(DA)_SingleLeafDamper_Square_ABK_Arctic	ABK_200x200	ES.11	Air Ducts equipment
Air Ducts equipment	(DA)_SingleLeafDamper_Square_ABK_Arctic	ABK_250x200	ES.11	Air Ducts equipment
Air Ducts equipment	(DA)_SingleLeafDamper_Square_ABK_Arctic	ABK_250x250	ES.11	Air Ducts equipment
Air Ducts equipment	(DA)_SingleLeafDamper_Square_ABK_Arctic	ABK_250x400	ES.11	Air Ducts equipment
Air Ducts equipment	(DA)_SingleLeafDamper_Square_ABK_Arctic	ABK_250x500	ES.11	Air Ducts equipment
Air Ducts equipment	(DA)_SingleLeafDamper_Square_ABK_Arctic	ABK_300x200	ES.11	Air Ducts equipment
Air Ducts equipment	(DA)_SingleLeafDamper_Square_ABK_Arctic	ABK_300x250	ES.11	Air Ducts equipment
Air Ducts equipment	(DA)_SingleLeafDamper_Square_ABK_Arctic	ABK_300x300	ES.11	Air Ducts equipment
Air Ducts equipment	(DA)_SingleLeafDamper_Square_ABK_Arctic	ABK_400x250	ES.11	Air Ducts equipment
Air Ducts equipment	(DA)_SingleLeafDamper_Square_ABK_Arctic	ABK_500x250	ES.11	Air Ducts equipment
Air Ducts equipment	(DA)_SingleLeafDamper_Square_ABK_Arctic	ABK_800x300	ES.11	Air Ducts equipment
Air diffusers	4АПН+3КСД М	4АПН 300x300+3КСД М	ES.12	Air diffusers
Air diffusers	4АПН+3КСД М	4АПН 450x450+3КСД М	ES.12	Air diffusers
Air diffusers	4АПН+3КСД М	4АПН 600x600+3КСД М	ES.12	Air diffusers
Air diffusers	4АПН-П+3КСД М	4АПН 300x300+3КСД М	ES.12	Air diffusers
Air diffusers	4АПН-П+3КСД М	4АПН 450x450+3КСД М	ES.12	Air diffusers
Air diffusers	4АПН-П+3КСД М	4АПН 600x600+3КСД М	ES.12	Air diffusers
Air diffusers	4АПН-П+3КСД М С	4АПН 300x300+3КСД М С	ES.12	Air diffusers
Air diffusers	4АПН_(перемок)	4АПН 450x450	ES.12	Air diffusers
Air diffusers	4АПН_(перемок)	4АПН 600x600	ES.12	Air diffusers
Equipment	42NL_NH_4_LEFT	42NL_NH_4_LEFT	ES.15	Air condition equipment
Fitting for Air Ducts	ADSK_AirDuctBranch_Square	Standart	ES.10	Fitting for square air ducts
Fitting for Air Ducts	ADSK_AirDuctBranch_Round	Standart	ES.09	Fitting for round air ducts
Fitting for Air Ducts	ADSK_AirDuctPlug_Round	ВСН	ES.09	Fitting for round air ducts
Fitting for Air Ducts	ADSK_AirDuctPlug_Square	Standart	ES.10	Fitting for square air ducts

Fig. 3. Example of marking up elements of BIM models using the specification.

Exporting Marked-Up BIM Models from RVT Format to IFC Format. "The Industry Foundation Classes (IFC) format is a platform–neutral specification of an open file

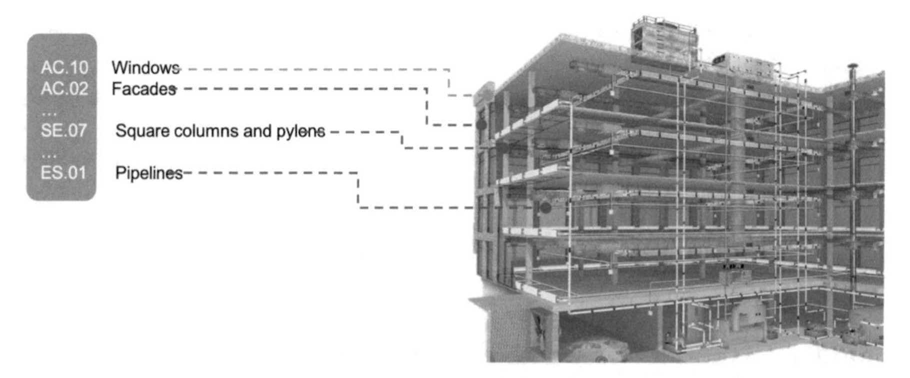

Fig. 4. Example of marking up building elements with a classifier code.

format that is not controlled by a single vendor or a group of vendors. It is an object-oriented data model file format developed by buildingSMART (until 2005, the International Alliance for Interaction, IAI) to facilitate interaction in the field of architecture, design and construction (AEC), and is a widely used format for cooperation in the field of building information. Projects based on modeling (BIM). The IFC model specification is open and available. It is registered by ISO and is the official international standard ISO 16739-1:2018. The IFC format is designed to describe data on architecture, design and construction and was created on the initiative of Autodesk" [13].

The export is performed in order to transform meta information about each element into a machine-readable text format.

Exporting Data to JSON Format. The JSON (JavaScript Object Notation) format is a standard text format for storing and transmitting structured data. Export to this format in order to transfer meta-information directly for analysis and learning of an AI model based on TF-IDF and logistic regression in text data format.

Clearing JSON Format from Unused Objects. When exporting from RVT format to IFC format, a description of elements that are not involved in the formation of building elements was uploaded. Since these elements have no value in the target attribute for analysis "Assembly code", to exclude erroneous and ambiguous classification, data about these elements were cleared from text files. Examples of such elements can be elements with attribute values "Category" = "IfcBuilding", "IfcBuildingStorey", "IfcOpeningElement" and others.

Normalization and Cleaning of Unused Text Data. Since it was revealed by experts that in order to classify elements, it is necessary to analyze the values of 4 attributes: Id, Category, Family, Type. Then, for further learning, it is also necessary to know information about the attribute where the data on the classification of the element is recorded.

Therefore, for subsequent work, we save the following data:

- Id. In JSON format it is written as "ID". Example attribute value: 86525;
- Category. In JSON format it is written as "Category". Example attribute value: IfcBuildingElementProxy;

- Family. In JSON format it is written as “Name”. Example of the attribute value: ADSK_Designer curtain_Water;
- Type. In JSON format, it is written as “Type”. Example of attribute value: KAC-ADIS25W:8313468;
- Assembly code. In JSON format, it is written as “Assembly code”. Example of an attribute value: ES.16.

There are also additional procedures in data normalization:

- Tokenization of words. Example: the value “ADSK_Designer curtain_Water” after tokenization will take the form “ADSK”, “Designer” “Curtain”, “Water”;
- Lowercase translation, removal of stop words and punctuation marks;
- Lemmatization. Words should be reduced to the initial (dictionary) form or lemma. Example: the values of “Pipe”, “pipes” are lemmatized into “pipe”.

Next, the cleared data is grouped by the values of the “Id” attribute in order to be able to analyze the data grouped element by element.

4.2 Development of an AI Model Based on TF-IDF and Logistic Regression

Having received the prepared data about each element, we proceed to analyze the information using the TF-IDF method to obtain the most important words-tokens that characterize each classifier.

To begin with, let’s analyze the distribution of classes in the context of elements in the information about the elements of the BIM models of the facility «Hotel» (see Fig. 5).

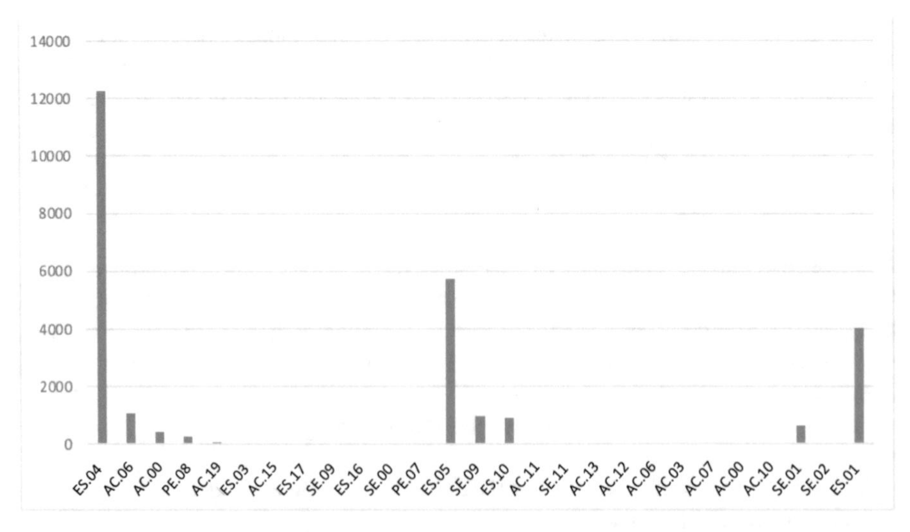

Fig. 5. Diagram with the distribution of the number of elements by classifier code.

The following conclusions can be drawn (see Fig. 6):

- In BIM models there are elements with 60 classifier positions out of 72. Therefore, there are 12 positions in which the AI model will not be able to learn, having examples of the classification of one project. However, more classes will be covered.
- There are 8 rare classes. The rarity is determined by the frequency of occurrence of elements with such a classifier code in the learning sample. If an element of a certain class has less than 300 examples, then such a class is considered rare.

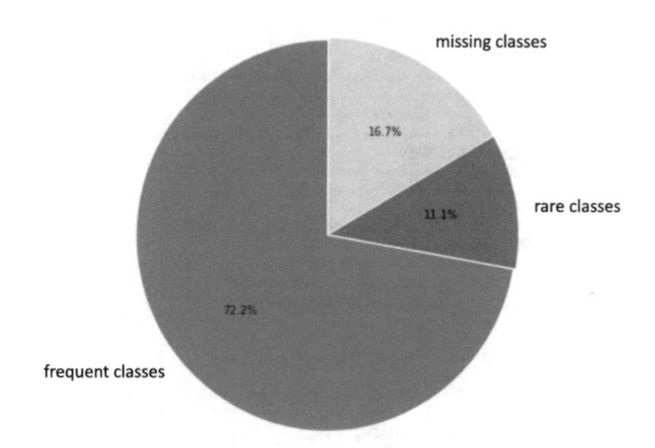

Fig. 6. Analysis of the use of classifier codes in elements.

After analyzing the information using the TF-IDF method, the most important words and phrases were identified. Figure 7 shows a list of these tokens and their frequency of occurrence in all documents submitted for analysis.

For further learning of the AI model, before using the logistic regression method, it is necessary to divide the total number of elements in the documents into two parts. On 80% (425 919 units) of the elements, we will create a learning sample, learn our AI model and use them for the logistic regression method. And 20% (106,480 pieces) of the elements are needed to create a test sample. At the same time, it is necessary to select random elements of different classes.

Then, using logistic regression, an AI model was created that could classify elements based on calculated features. A link was built between the token defining the element and its class (Table 1).

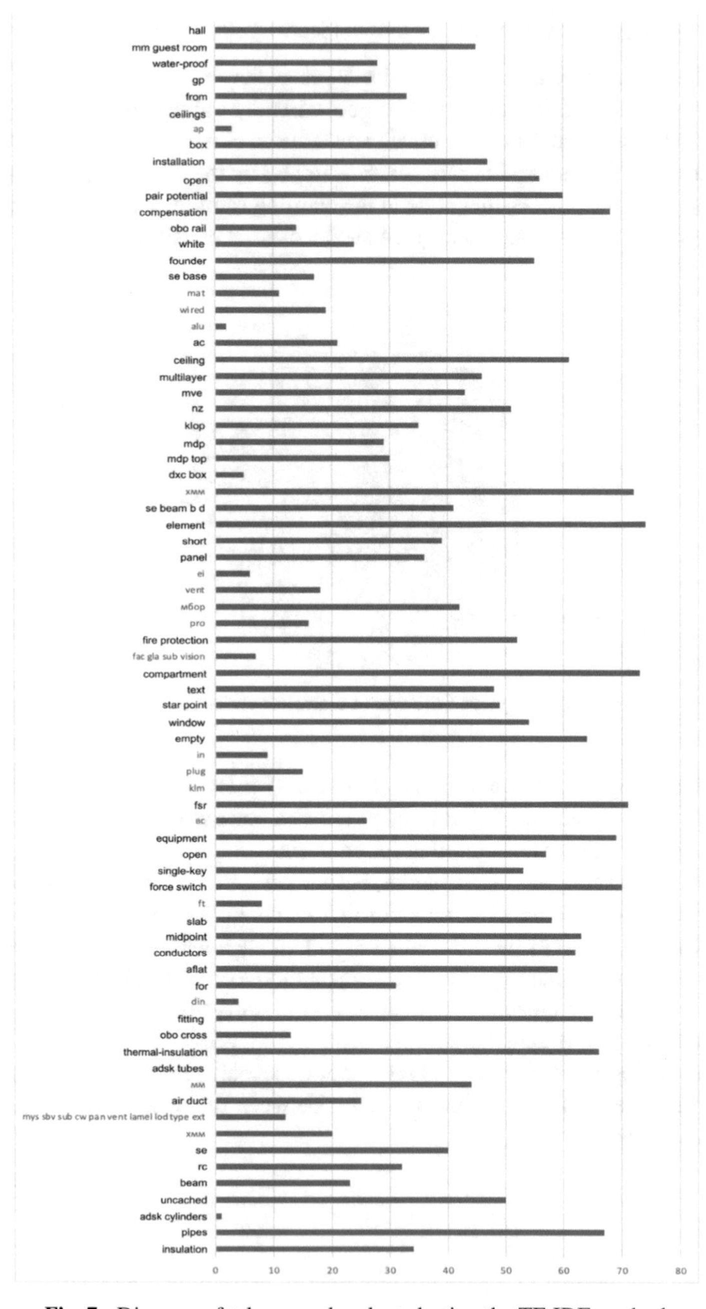

Fig. 7. Diagram of token words selected using the TF-IDF method.

Table 1. Example of a combination of a token word and a classifier code.

Token word	Assembly code	Description of assembly code
Beam	SE.09	Beams
Isolation	ES.04	Insulating air ducts
Plate	SE.13	Structural units
Single-key switch	PE.05	Sockets, switches
Pipes	ES.01	Pipelines
Ceiling	AC.06	Ceilings

Next, the learned AI model was tested on a test sample of 20% of the elements (106 480 units). Having received the following classification results (Table 2):

Table 2. Results of the AI model on a test sample.

Criteria	Value
Precision	0.99
F1 (macro)	0.92
F1 (weighted)	0.99

The classification results on a test sample of the same object on which the AI model was learned are quite high, however, on the data of a new object on which the AI model was not learned, the indicators decreased.

The new data on the "Company's office" facility were also divided into two parts. At the first iteration, 80% of the elements were tested, and the remaining 20% of the elements were further divided into several parts of 300 elements in order to determine the number of iterations for which the AI model will be able to show the percentage of correct predictions (Precision) equal to 0.86.

The results presented in Table 3 show that precision does not increase much after the third iteration, which means that after the third iteration we received a ready-made working AI model (see Fig. 8, 9 and 10).

Table 3. Results of the AI model with the process of active learning at a new facility.

Criteria	I iteration 80% of the elements	II iteration +300 elements	II iteration +300 elements	IV iteration +300 elements
Precision	0.72	0.76	0.86	0.87
F1 (macro)	0.29	0.35	0.44	0.44
F1 (weighted)	0.75	0.82	0.83	0.82

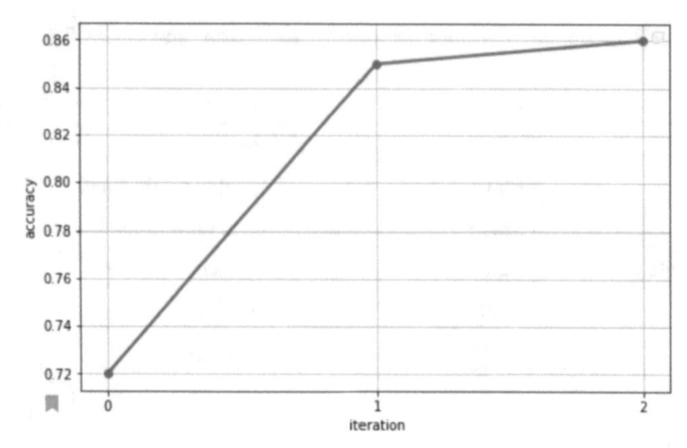

Fig. 8. The growth of the precision (Accuracy) index with active lerarning by the number of iterations.

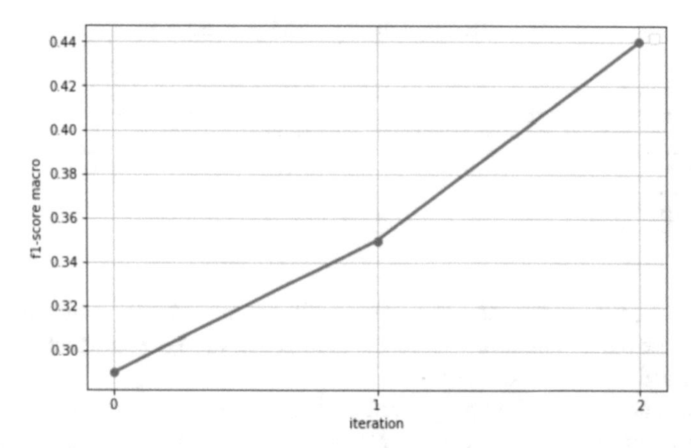

Fig. 9. The growth of the F1 (macro) indicator with active learning by the number of iterations.

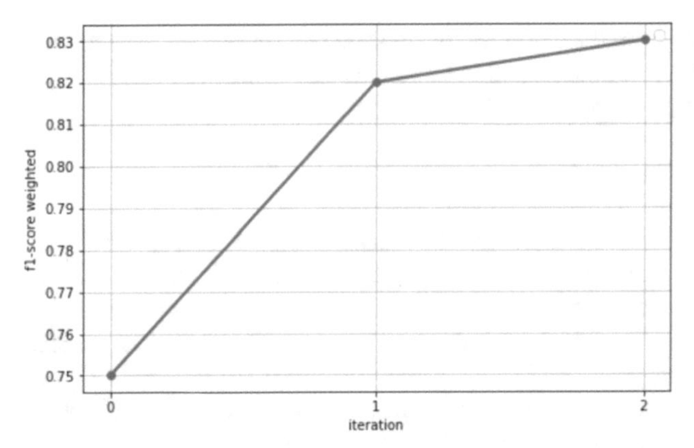

Fig. 10. The growth of the F1 (weighted) indicator with active learning by the number of iterations.

5 Results

As a result of the conducted research, it was possible to answer the following questions asked at the beginning.

Based on the transferred 45 units of BIM models of the «Hotel» facility with a total number of elements of 532 399 units, the classification code was marked into each element of the BIM model. Temporary labor costs for marking - 5 days. In the markup results, the minimum data set for determining the classification of an element was determined by expert means: Id, Category, Family, Type.

Next, the main stages of data normalization were determined before creating an AI model:

- Export of data in machine-readable text form;
- Cleaning of unused data and objects;
- Tokenization of words, lowercase translation, removal of stop words and punctuation marks, lemmatization.

The next step was to determine the most significant words using the TF-IDF method. After that, the bundle "Word" -> "Assembly code" was carried out.

Further, the elements of the "Hotel" object were divided into a training and test sample in a percentage ratio of 80/20, respectively. After that, on the text sample, we got a fairly high precision of the forecast, equal to 0.99.

Next, the elements of the new facility "Company's office" were taken. As expected, the accuracy of predictions will decrease on a new object unfamiliar to the AI model, but the goal was to determine for how many iterations of retraining the Accuracy indicator will be at least 0,86, at which the AI model becomes effective. The number of iterations is 3.

Recommendations:

- In order for the accuracy of the AI model not to have sharp jumps at the beginning of work at a new facility, it is necessary to train and retrain at more than 3 objects of different categories: industrial, civil, linear and other;

- All classes from the classifier list must be present in these objects. At the same time, each code must be repeated in more than 300 elements;
- Sufficient precision of the AI model - 0.86.

6 Discussion

There are difficulties that do not allow achieving high quality criteria of the AI model. For example, training a model on the CCI classifier, which has more than 138 million positions, should. More than 41 billion objects should be placed to start training. The approach to creating AI models described in this study may not be suitable for solving such a problem.

Also, for a more accurate prediction of new construction projects, models of different types of construction objects should be marked up.

7 Conclusion

Analyzing the time spent on marking up and training the AI model (about a month), we can conclude that the manual and semi-automatic method will not achieve the same results in terms of speed and accuracy of classification. On average, it takes about 9 h for the classifier to mark up elements of one BIM model, while during the entire design period it is necessary to check, correct, fill in and keep the data up to date. With the automated method using the classification by the AI model, the labor costs turned out to be about 10 min for each BIM models.

References

1. Meisels, M., et al.: Engineering and construction industry outlook. Deloitte **11** (2021)
2. Gusev, A.V., Dobridnyuk, S.L.: Artificial intelligence in medicine and healthcare. Inf. Soc. **4–5**, 78–93 (2017)
3. Solihin, W., Eastman, C.: Classification of rules for automated BIM rule checking development. Autom. Constr. **53**, 69–82 (2015)
4. Petrochenko, M.V., Nedviga, P.N., Kukina, A.A., Sherstyuk, V.V.: Classification of construction information in BIM using artificial intelligence algorithms. Bull. MGSU **17**(11), 1537–1550 (2022)
5. Afsari, K., Eastman, C.: A comparison of construction classification systems used for classifying building product models. In: Conference: 52nd ASC Annual International Conference Proceedings (2016)
6. Machado, M.R., Karray, S., de Sousa, I.T.: Light GBM: an effective decision tree gradient boosting method to predict customer loyalty in the finance industry. In: 14th International Conference on Computer Science Education, pp. 1111–1116 (2019)
7. Chen, T., Guestrin, C.: XGBoost: a scalable tree boosting system. In: The 22nd ACM SIGKDD International Conference on Knowledge Discovery and Data Mining. Association for Computing Machinery, New York (2016)
8. Liu, X., Wang, T.: Application of XGBOOST model on potential 5G mobile users forecast. In: Sun, J., Wang, Y., Huo, M., Xu, L. (eds.) Signal and Information Processing, Networking and Computers. LNEE, vol. 917, pp. 1492–1500. Springer, Singapore (2023). https://doi.org/10.1007/978-981-19-3387-5_177

9. Catboost. https://catboost.ai/en/docs/concepts/python-reference_catboostclassifier. Accessed 15 Oct 2023
10. Creating an ICD-10 code prediction model based on the text of the disease description. https://habr.com/ru/articles/673312/. Accessed 15 Oct 2023
11. Uhanov, N.: Master's dissertation. Application of the text information analysis method for the problem of classification of scientific articles. SFU, Krasnoyarsk (2021)
12. Natural Language Processing (NLP) with Machine Learning in Python. https://habr.com/ru/companies/otus/articles/687796/. Accessed 15 Oct 2023
13. Engineering modeling 3D-BIM model. https://3dbim.pro/news/23/. Accessed 15 Oct 2023

Computer Vision System for Monitoring User Attention in Interactive User Interfaces

Anton Ivaschenko[1] (✉), Vladimir Avsievich[1], Margarita Aleksandrova[2], Ivan Legkov[2], and Kirill Sheshulin[2]

[1] Samara State Medical University, 89 Chapaevskaya, Samara, Russia
anton.ivashenko@gmail.com

[2] SEC "Open Code", 55 Yarmarochnaya, Samara, Russia

Abstract. The paper presents a new solution of computer-human interfaces personalization based on external analysis of the users' activity based on face recognition. The approach includes tracking of the human face parts and emotion recognition using an external video camera integrated with an intelligent computer vision system. The knowledge base is used to compare the data from a computer vision system and action tracker and process it in a recognizer component. In case the identified activity (like emotion of number of actions) differs from the target one there are generated virtual stimuli that attract the user attention and provide the required activity change. The proposed approach was used in a medical system for psychological testing and also for proctoring in a digital educational platform. Testing experiments included 5000 images and proved a possibility to fairly clearly distinguish between attentive users and users being distracted. Both applications gathering an independent feedback form users in order to control the correct system functioning and adapt the components of user interface for user attraction when needed.

Keywords: Computer Vision · Eye Tracking · Personalization · Adaptive User Interfaces · Use Attention Analysis

1 Introduction

Modern computer vision technologies play an important role in the field of facial recognition and extracting key points from images. They provide capabilities for analyzing unique facial features, which has applications in various fields including security, authentication and behavior analysis. An important area of their application is obtaining feedback from users of interactive computer interfaces, allowing them to monitor and control their attention and involvement.

Video analytics systems, working with high-definition video cameras, are able to collect detailed data about faces. They allow you to highlight facial features, such as contours, shape of eyes, nose, mouth and other characteristics. This data can be further used to determine unique biometric parameters that allow each individual to be uniquely identified. Video camera being installed in front of the person tested, allows you to

A. Gibadullin (Ed.): DITEM 2023, LNNS 942, pp. 206–217, 2024.
https://doi.org/10.1007/978-3-031-55349-3_17

obtain a detailed image of the participant in the process. Computer vision algorithms can recognize faces with high accuracy and also extract key points to create unique biometric characteristics.

An important element of control is the analysis of the participants' behavior. By formalizing the context of object activity, it is possible to identify non-standard actions that go beyond established patterns. For example, the system can detect suspicious movements or attempts at fraud.

In this paper there is presented a new approach for computer vision system, which provides monitoring user attention in interactive user interfaces.

2 State of the Art

Implementing automated analysis of user activity is an important step in improving the efficiency of modern user interfaces [1, 2]. The main problem being currently solved is related to the need to control and ensure high user engagement with a constant information load and the presence of distractions. To solve this, various technologies can be used, for example, logging user actions with subsequent analysis of actions performed in the user interface. However, this approach significantly depends on the implementation features of the user interface and the ease of its use.

A promising tool for solving this problem is external control based on the use of video analytics tools for facial recognition [3–6]. Most methods are based on implementation of artificial neural networks, which provide a tool for face and emotion recognition [7, 8], head and eyes activity analysis, and thus give sufficient data to analyze user activity and pay attention to changes in the user interface.

Integration of modern computer vision technologies, neural networks and behavior analysis models into video analytics systems provides reliable tools for facial recognition, extracting key points from images and comparing characteristics. These tools are used in various fields that require accurate identification of individuals and analysis of their actions. Additionally, social aspect analysis can be applied to identify anomalous behavior among participants. This includes analyzing facial expressions, gestures, and other signs that may indicate stress or intentional deception.

The use of these technologies allows us to study the peculiarities of perception of user interface components in human-computer interaction systems [9, 10]. This is especially challenging when introducing virtual and augmented reality interfaces, as a result of which it is necessary to achieve a highly immersive effect [11, 12], which is in demand in medical and psychological diagnostics and rehabilitation applications, as well as digital education.

The immersion control system can be built using eye-tracking sensors [13, 14] or a computer vision system for face recognition. The proposed below solution architecture and experiment follow the logic of accented visualization method implementation in interactive computer-human user interfaces [15, 16].

Analysis of known publications and developments in the field of application of modern user interfaces, as well as face recognition systems, made it possible to identify a certain gap. On the one hand, modern computer vision systems make it possible to quite accurately determine the user's state and its changes considering the human factor.

On the other hand, these capabilities are poorly used to control feedback in complex and interactive user interfaces of modern applications, especially during their operation in real time. This provides a wide scope of work to eliminate this gap as part of the personalization of modern interactive applications.

3 Materials and Methods

The proposed solution architecture is presented in Fig. 1. User activity is tracked by monitoring facial actions using an external video camera integrated with a computer vision system. User activity patterns and the logic of their processing are captured in the knowledge base, which is implemented in the form of Ontology. The knowledge base is used to compare the data from a computer vision system and action tracker and process it in a recognizer component. In case the identified activity (like emotion of number of actions) differs from the target one there are generated virtual stimuli that attract the user attention to certain areas and object in virtual space and provide the required activity change.

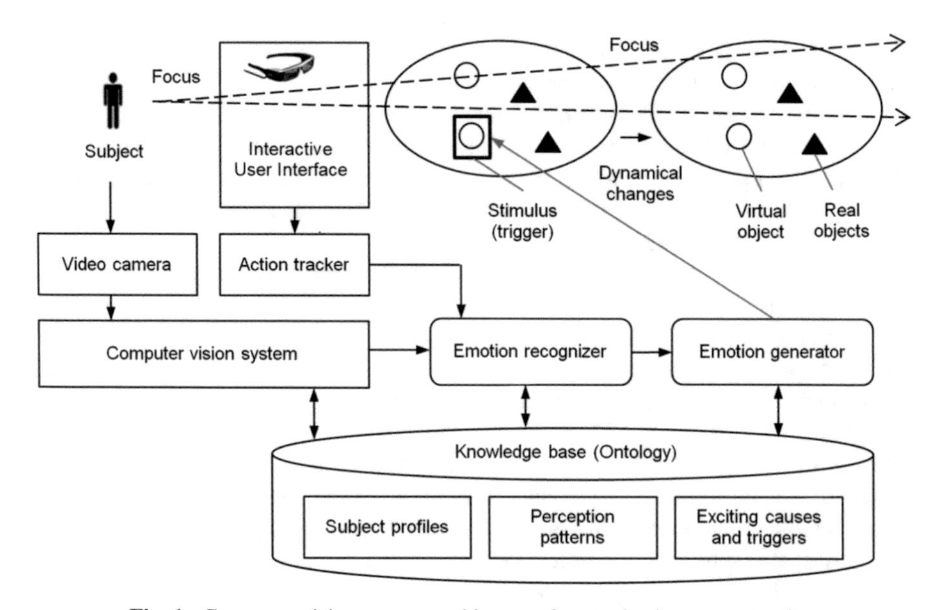

Fig. 1. Computer vision system architecture for monitoring user attention.

Determining the user's attention and involvement is based on monitoring facial expressions, emotions, eye sight and focus direction and the number of head movements. The combination of these parameters in dynamics allows you to track how much the user's attention changes over time. The basis for determining these parameters is facial recognition.

The input image is represented as a three-dimensional array of pixels. The process begins by applying a convolutional neural network, which scans the image using various filters to extract important features.

Convolutional layers are the basis of convolutional neural networks because they contain the training weights that extract the features needed to classify images. The convolution kernel is a two-dimensional matrix of coefficients. This kernel is applied to the image region and the dot product between the input pixels and the filter is then calculated. The kernel size is usually smaller than the image dimensions:

$$(f * g)[m, n] = \sum_{k,l} f[m-k, n-l] * g[k, l], \quad (1)$$

where f is the original image matrix, and g – convolution kernel.

The final result of a series of dot products of the input data and the filter is known as a feature map. The weights in the kernel remain fixed as it moves across the image, also known as parameter sharing.

The size of the feature map can be calculated using the formula:

$$H_2 = \frac{(H_1 - F + 2 * P)}{S + 1}, \quad W_2 = \frac{(W_1 - F + 2 * P)}{S + 1}, \quad (2)$$

where H_1 is the height of the input image; W_1 – input image width;

H_2 – height of the feature map; W_2 – width of the feature map;

F – convolution kernel size; P – number of pixels filled with zeros;

S – step of the convolution kernel.

Applying nonlinear activation functions (such as ReLU) adds nonlinearity to the model, allowing it to learn complex dependencies in the data. Rectified linear activation function is a type of activation function that is linear in the positive dimension but equal to zero in the negative dimension. The break of a function is a source of nonlinearity.

The rectified linear function is expressed by the formula:

$$f(x) = \max(0, x), \quad (3)$$

where x is a real number.

This function has a number of advantages such as simplicity and computational efficiency, overcoming the gradient decay problem, overcoming the problem of gradient growth, and networks with ReLU can be deeper:

Subsampling layers reduce the dimensionality of data, preserving important features and reducing computational load. If some features have already been identified during the previous convolution operation, then such a detailed image is no longer needed for further processing, and it is compressed to a less detailed one. In addition, filtering out unnecessary details helps avoid overtraining.

This layer can be described by the formula:

$$x^l = f\left(a^l * subsample\left(x^{l-1}\right) + b^l\right), \quad (4)$$

where x^l is the output of layer l; f – activation function; a^l, b^l – layer shift coefficients;

subsample – operation of sampling local maximum values.

The resulting features are processed using fully connected layers for further classification. This layer models a complex nonlinear function, optimizing which improves the quality of recognition.

The calculation of neuron values can be described by the formula:

$$x_j^l = f\left(\sum_i x_i^{l-1} * w_{i,j}^{l-1} + b_j^{l-1}\right), \tag{5}$$

where x_j^l is the feature map j (output of layer l); f – activation function;

b_j^l – layer shift coefficients l; $w_{i,j}^l$ – matrix of weight coefficients of layer l;

Using a threshold function allows you to determine whether an image region contains a face. At the stage of face localization, if a threshold function is applied, regions corresponding to faces can be detected and highlighted. Choosing a MMOD Human model Face Detector was driven by its effectiveness in face recognition tasks. The model was initialized with pre-trained weights, allowing it to use previous knowledge to speed up learning. Fine-tuning of the hyperparameters was done taking into account learning rate, batch size, and momentum. This process involved an iterative cycle of experimentation, testing, and refinement.

Training was characterized by iterative epochs, during which batches of augmented data were fed into the model. Back propagation and gradient descent methods were used to minimize the detection error. In addition, an adaptive learning rate scheduling function was implemented to dynamically adjust the learning rate. To prevent model overload, an early stopping mechanism was built in to evaluate its performance on a separate set. Model checkpoints are saved at regular intervals, recording the state of the model.

To

extract key points from a face image we used a shape_predictor_68_face_landmarks model (see Fig. 2). This model contains information about which facial points need to be extracted to form 68 key points, including points on the eyes, nose, mouth and cheeks. The resulting coordinates can be used for further analysis or visualization of key points in the image.

To extract a vector of 128 values into which the representation of a face is encoded, a special method is used, passing an image with a face into it. This method parses the image and returns an array representation of the face. This array is a numeric vector that uniquely represents a specific person. This vector can be used to compare faces and determine their similarity.

To extract a vector representation of a face, a pre-trained ResNet-50 model, a face image, and pre-obtained key points on the face are used. If you have multiple encoded faces, you can compare them using various similarity metrics, such as Euclidean distance.

To solve the problem of detecting head rotation, an algorithm based on extracting key points of the face (landmarks) is used. These key points represent objects on the face such as eyes, nose, mouth, etc. They can be used to analyze head orientation.

The head rotation assessment process includes the following steps. Facial keypoints represent the coordinates of anatomical features of the face, such as the corners of the eyes, nose, mouth, etc. Using the location of key points, we can calculate head rotation angles in 3D space (yaw, pitch, roll). These angles are the horizontal, vertical and longitudinal rotation of the head. The library calculates a 3x3 rotation matrix R which represents the transformation to the image coordinate system.

The resulting angles can be used to analyze the position and orientation of the head in the image. For example, the yaw angle indicates the direction in which the head is

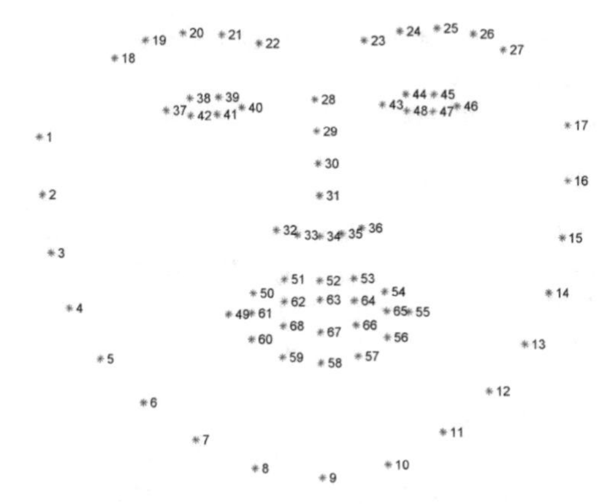

Fig. 2. Marking key points on the face.

turned left or right, the pitch angle indicates an upward or downward tilt, and the roll angle indicates a sideways tilt.

4 Results

The proposed system was implemented using a configuration of a convolutional neural network from the dlib library. This is the optimal solution for real-time work, which makes it possible to accurately process input data from the video stream and perform assigned tasks.

In order to automate the detection and classification of images, the FaceRecognition library was used, which is a library for face recognition. The advantages of using this library include the following: one of the fastest libraries; MIT Open License; access from Python, C++, C # languages; Linux and Windows operating systems supporting; OpenCV library of computer vision algorithms was used.

To obtain the necessary weights, the model was trained on 5,000 images, divided into a training set and a test set of 80% by 20%. Before the training process began, careful preparations were made to optimize the data set and ensure that the model responded to the training regime.

The following key procedures were performed. Each image in the dataset was subjected to uniform resizing and cropping procedures to establish a uniform scale, ensuring equal representation of all samples. In addition, careful alignment of facial features was performed to center the face in each image, ensuring a standardized perspective.

To reduce the possible influence of illumination unevenness, an illumination normalization process was used. This made it possible to unify the light levels in the data set and thereby reduce their impact on the model training process.

Accurate localization of critical facial landmarks such as eyes, nose, and mouth has been achieved using advanced landmark detection techniques. This allowed the model to focus on key areas to accurately estimate head position. A stringent quality control procedure was implemented to identify and eliminate any anomalies or incorrect distributions in the data set. Outliers or questionable samples were carefully screened to help maintain the integrity of the data set.

After preparation and refinement of the data set, the training phase began, which included a number of targeted strategies and techniques aimed at optimizing the performance of the model. Hyperparameters such as learning speed and batch size were fine-tuned. This process allowed us to achieve an optimal balance between computational efficiency and model accuracy. By integrating transfer learning, the model was endowed with existing knowledge, allowing it to gain a basic understanding of facial features and head position dynamics. This allowed us to speed up the model training process.

The learning process was characterized by iterative epochs where batches of data were systematically presented to the model to refine the internal weights. An early termination mechanism was built in to prevent overkill and thus ensure optimal model convergence. By following these structured procedures, the model was systematically refined to a state of optimized proficiency, allowing it to solve complex head pose estimation problems in real-world environments with increased accuracy and efficiency.

This process allows you to analyze the orientation of the head in the image using key points of the face. FaceRecognitionDotNet provides convenient means to extract and analyze these key points, making head rotation assessment accessible and convenient.

5 Discussion

The result of the algorithm for detecting faces and key points on test subjects (see Fig. 3). The proposed approach was used in a medical system for psychological testing and also for proctoring in digital educational platform. Both applications gathering an independent feedback form users in order to control the correct system functioning and adapt the components of user interface for user attraction when needed.

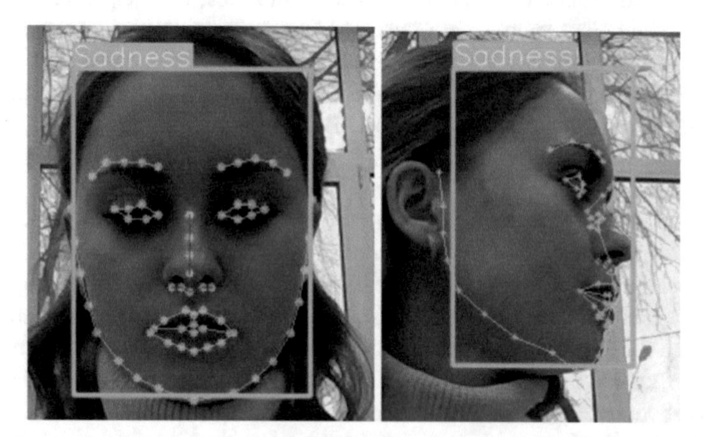

Fig. 3. Detection of a face, emotion and key points when turning the head.

As you can see from the images, turning the head to the side when part of the face is not in the camera's field of view does not interfere with face recognition.

Let's compare the results of head rotation detection and the ability to track this moment on video. Two videos were selected in which two people take a test, but during the test, in one of the videos the test taker turns away. The video data was processed by the algorithm and graphs were constructed using the resulting rotation angles; data normalization was previously performed.

Activity tracking graphs make it possible to fairly clearly distinguish between an attentive user (see Fig. 4) and a user who was distracted during the experiment (see Fig. 5).

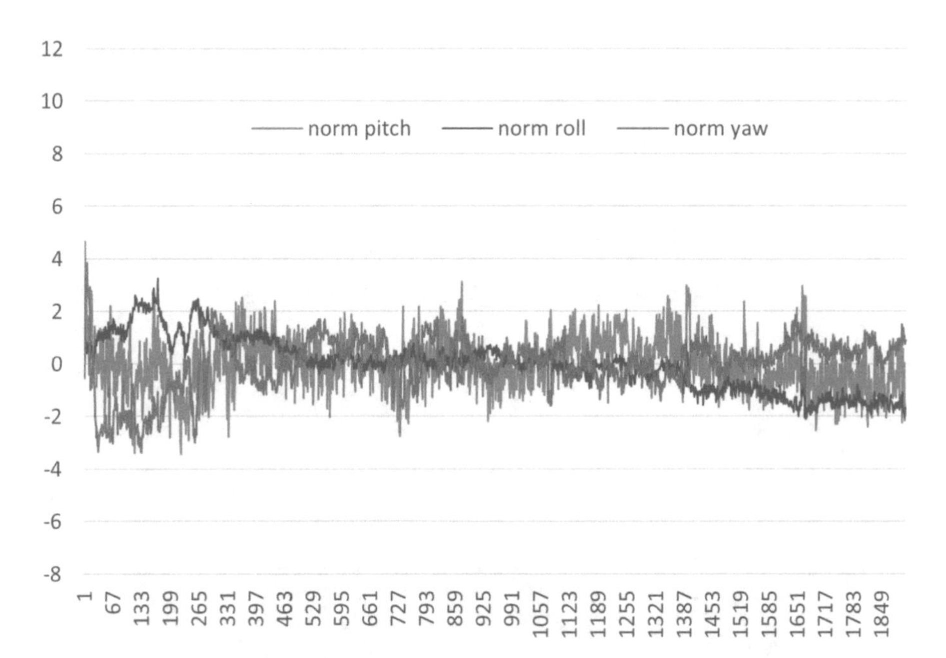

Fig. 4. Head rotation angle diagram for an attentive user.

As you can see in Fig. 5, there are spikes in the rotation data, which signals the head turns of the second participant. Let's arrange these diagrams into corners and compare them with each other. Let's compare the graphs of the two test subjects, where the orange line marks the graph of a person who was not distracted, and the blue line who was distracted at a certain point in time (see Fig. 6, 7 and 8).

Analyzing the diagram data, we can conclude that one of the users turns his head away from the monitor along the longitudinal and vertical axis in the interval from frame 1366 to frame 1496 and from frame 1886 to frame 1970, and from the data obtained we can also draw a conclusion about the direction of head rotation.

It should be noted that in user interface adaptation tasks, it is important not so much to identify differences between different users, but to identify changes in interest for one user. In this regard, interface personalization is associated with the analysis of relative changes in the user's behavior and emotional background.

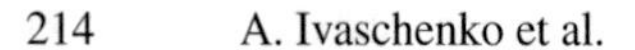

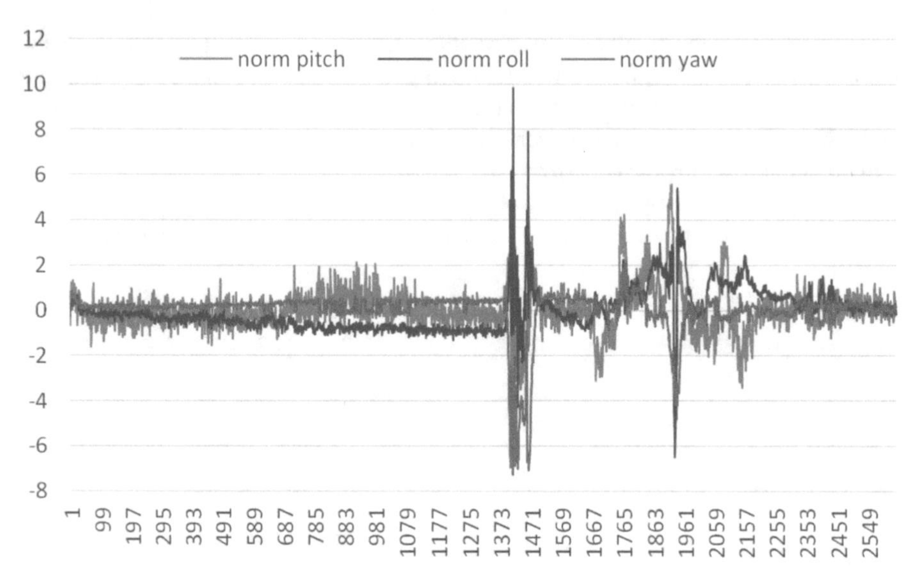

Fig. 5. Head rotation angle diagram for destruction.

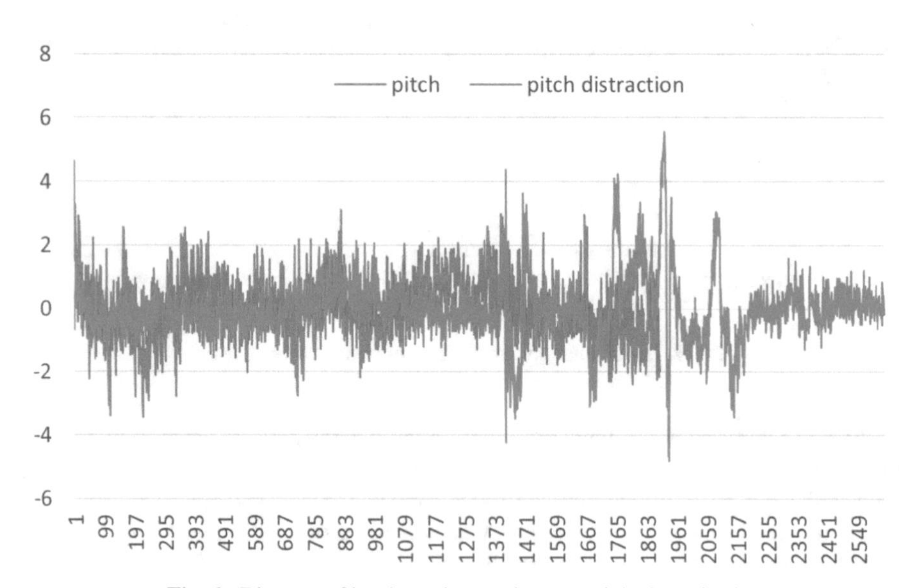

Fig. 6. Diagram of head rotation angles around the lateral axis.

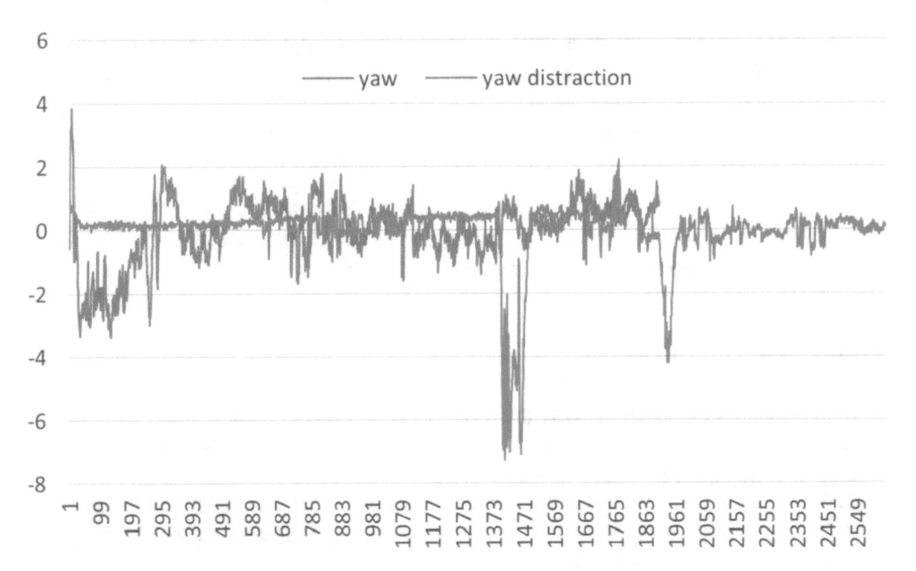

Fig. 7. Diagram of head rotation angles around the vertical axis.

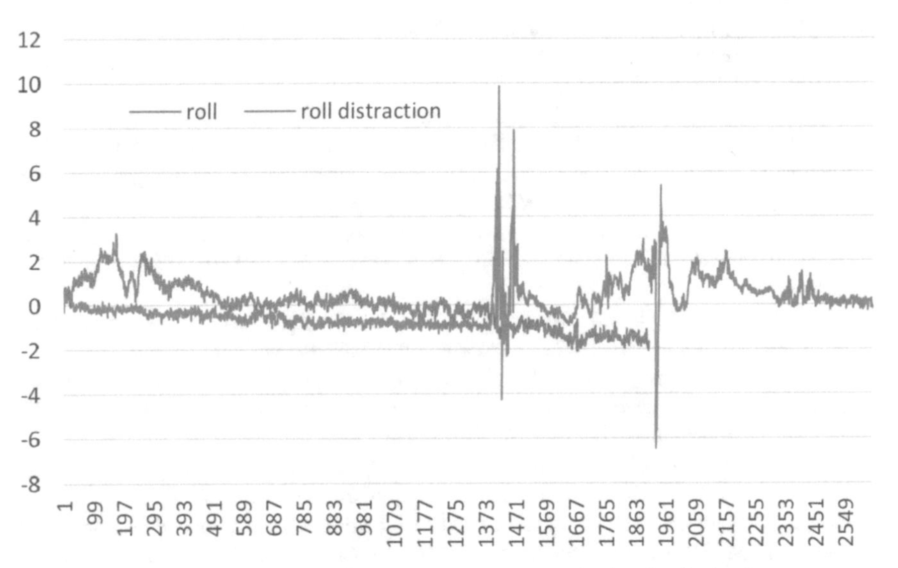

Fig. 8. Diagram of head rotation angles around the longitudinal axis.

6 Conclusion

Adaptive user interfaces are natural and promising to improve computer-human interfaces at the modern level of information technologies application. Face recognition and emotions analysis allows to improve the efficiency and ergonomics of modern user interfaces making them more attractive and personalized for different users.

The basic areas of application of the proposed approach include physiological diagnostics and rehabilitation, digital education and intelligent production. All these problem domains require improving the logic and technology of computer-human interaction considering the new possibilities of user interfaces capable of building the immersive reality.

Next research steps include improving the technology of user activity analysis and studying the ways of their combination in order to provide high accuracy and performance of personalized user interfaces.

References

1. Thakur, N., Han, C.Y.: An activity analysis model for enhancing user experiences in affect aware systems. In: 2018 IEEE 5G World Forum (5GWF), pp. 516–519, Silicon Valley, CA, USA (2018). https://doi.org/10.1109/5GWF.2018.8517032
2. Park, So., Park, Su., Ma, K.: An automatic user activity analysis method for discovering latent requirements: usability issue detection on mobile applications. Sensors **18**, 2963 (2018). https://doi.org/10.3390/s18092963
3. Mohammad, S.M.: Facial recognition technology. SSRN Electron. J. (2020). https://doi.org/10.2139/ssrn.3622882
4. Oloyede, M., Hancke, G., Myburgh, H.: A review on face recognition systems: recent approaches and challenges. Multimed. Tools Appl. **79**, 27891–27922 (2020). https://doi.org/10.1007/s11042-020-09261-2
5. Singh, S., Prasad, S.V.A.V.: Techniques and challenges of face recognition: a critical review. Procedia Comput. Sci. **143**, 536–543 (2018). https://doi.org/10.1016/j.procs.2018.10.427
6. Remone, R., Dash, S.: Face recognition and face detection benefits and challenges. Eur. Chem. Bull. **12**, 2561–2566 (2023). https://doi.org/10.31838/ecb/2023.12.si6.226
7. Badrulhisham, N., Mangshor, N.: Emotion recognition using convolutional neural network (CNN). J. Phys. Conf. Ser. **1962**, 012040 (2021). https://doi.org/10.1088/1742-6596/1962/1/012040
8. Lu, X.: Deep learning based emotion recognition and visualization of figural representation. Front. Psychol. **12**, 818833 (2022). https://doi.org/10.3389/fpsyg.2021.818833
9. Mohammed, Y.B., Karagozlu, D.: A review of human-computer interaction design approaches towards information systems development. Brain Broad Res. Artif. Intell. Neurosci. **12**, 229–250 (2021). https://doi.org/10.18662/brain/12.1/180
10. Cioczek, M., Czarnota, T., Szymczyk, T.: Analysis of modern human-computer interfaces. J. Comput. Sci. Inst. **18**, 22–29 (2021). https://doi.org/10.35784/jcsi.2403
11. Barton, A.C., Sheen, J., Byrne, L.K.: Immediate attention enhancement and restoration from interactive and immersive technologies: a scoping review. Front. Psychol. **11**, 2050 (2020). https://doi.org/10.3389/fpsyg.2020.02050
12. Fathutdinova, K., Vulfin, A., Vasilyev, V., Kirillova, A., Nikonov, A.: Psycho-emotional state analysis system of APCS operator. In: Proceedings of ITNT 2020 - 6th IEEE International Conference on Information Technology and Nanotechnology, p. 9253187 (2020)
13. Jacob, R.J.K., Karn, K.S.: Eye tracking in human-computer interaction and usability research: ready to deliver the promises. The mind's eye: cognitive and applied aspects of eye movement research, pp. 573–605 (2003)
14. Duchowski, A.T.: A breadth-first survey of eye tracking applications. Behav. Res. Methods Instrum. Comput. (BRMIC) **34**(4), 455–470 (2002)

15. Ivaschenko, A., Sitnikov, P., Surnin, O.: Accented visualization for Augmented Reality. Emerging Topics and Questions in Infocommunication Technologies, pp. 74–97. Cambridge Scholars Publishing, Cambridge (2020)
16. Surnin, O., et al.: Augmented reality implementation for comfortable adaptation of disabled personnel to the production workplace. In: Proceedings of the 35th Annual European Simulation and Modelling Conference, pp. 64–69 (2021)

Processing of Data on Power Consumption in the Management of Energy Resources of the Regional Electrical Complex

Viktor Gnatuk[1], Oleg Kivchun[1], Dmitrii Morozov[1](✉), and Geetha Devi[2]

[1] Immanuel Kant Baltic Federal University, Kaliningrad, Russia
gnatukvi@mail.ru

[2] National University of Science and Technology, Muscat, Oman

Abstract. The article considers the theoretical justification of the stages of data processing on power consumption in the management of energy resources of the regional electrical complex. The development of modern technologies has made it possible to significantly increase the ability to measure parameters at electric power facilities, which is why it has become possible to accumulate large amounts of data using digital metering devices. Numerous studies of the structure and properties of the measurement data have led to the conclusion that they belong to non-Gaussian, i.e. the central limit theorems and the law of large numbers do not work on such data. Therefore, it is advisable to carry out their qualitative processing using the methods of rank analysis. The article proposes and theoretically substantiates an approach to creating a digital energy efficiency platform based on vector rank analysis in a computational and parametric complex. It will allow processing multidimensional arrays of non-Gaussian data and implementing scientific methods in order to obtain deterministic invariants for the decision-making preparation system. A distinctive feature of the digital energy efficiency platform is a one-parameter digital twin, the basis of which is an OLAP data cube by parameter, which allows long-term storage of arrays of non-Gaussian data of the regional electrical complex for electrical equipment used to generate deterministic invariants on the basis of which decisions are made.

Keywords: Processing · Data · Power Consumption · Management · Energy Resources · Regional Electrotechnical Complex · Digital Platform · Single-Parameter Digital Twin · OLAP Cube · Digital Profile

1 Introduction

The most important condition for the functioning of regional electrotechnical complexes (REC) at present is the necessary transition to digital energy. To this end, a large number of roadmaps, strategies, standards and other documents have been developed. In addition, new-generation digital devices are being introduced at all levels of REC management, which allow obtaining reliable data on energy consumption in real time, with high accuracy. The capabilities of new software and hardware complexes developed with the

A. Gibadullin (Ed.): DITEM 2023, LNNS 942, pp. 218–232, 2024.
https://doi.org/10.1007/978-3-031-55349-3_18

use of new digital technologies allow analyzing, processing and presenting the results of the functioning of all subsystems of the REC in order to improve the quality of energy management.

The main purpose of energy resources management is to ensure sustainable power consumption in various modes of operation of the REC. As a result of the installation of a large number of digital devices at energy facilities, it became possible to record significant amounts of measurements, which, in the future, can be converted into data. Based on the study of many samples of such data, it was found that they have a non-standard structure and are non-Gaussian [1–3]. Therefore, the most important process that improves the quality of energy management is the scientifically based processing of data obtained from measurement devices in real time in order to develop deterministic invariants on the basis of which control actions are formed. This process involves the study of all stages of the data life cycle, which are the stages of the process covering various data states, including their fixation, subsequent analytical processing, aggregation and transformation into control actions.

2 Calculation and Parametric Complex for Processing Data on Electricity Consumption

Professor V.I. Gnatyuk in [4] showed that the object of research of the REC is a computational parametric complex (RPC), which is formed by the results of measurements and fixed on a material carrier in a form suitable for analysis, processing, storage, transformation and transmission, an interconnected set of parametric data reflecting with quantitative, the qualitative, as well as the dynamic sides of the functional properties, both of the objects individually and of the REC as a whole. The RPC can be single-parameter or multi-parameter, depending on the number of measured parameters. Its main task is the formation of deterministic invariants, which in the future will become the main one for decision-making in the management of energy resources.

The implementation of the RPC can be carried out on a digital platform [5–7], the basis of which is a one-parameter digital double of the REC, which is a tuple of digital profiles constantly changing under the influence of the software functionality, limited by the number of transactions necessary to achieve the management goal. In relation to the PKK, there are single-parameter, multiparametric and nomenclatural-parametric digital doubles. The basis of the digital twin is an OLAP data cube by parameter - a multidimensional, long–term stored array of technocenosis data on power consumption, used in the process of interactive analysis of energy efficiency. The data included in it are continuously used in the energy management circuits and are updated in real time in the data warehouse [5–7]. This allows you to have the most up-to-date data and the most popular elements of the software functionality.

Figure 1 shows the structure of the process reflecting the different states of data on electricity consumption within the framework of the energy management process. In fact, their life cycle is shown, which includes the stages of fixation, subsequent analytical processing, aggregation and transformation into control actions.

The readings of devices of automated systems for monitoring and metering of electricity (AIS KUE) are processed in a measuring complex, which is understood as a combination of means, methods and auxiliary devices designed for measurement tasks. The operation of the measuring complex allows collecting and parameterizing information about the properties of the REC in real time. Quantitative description and careful systematization of information allow us to obtain numerical values of parameters and form a database. It should be noted that it is at this moment that information about REC facilities turns into data.

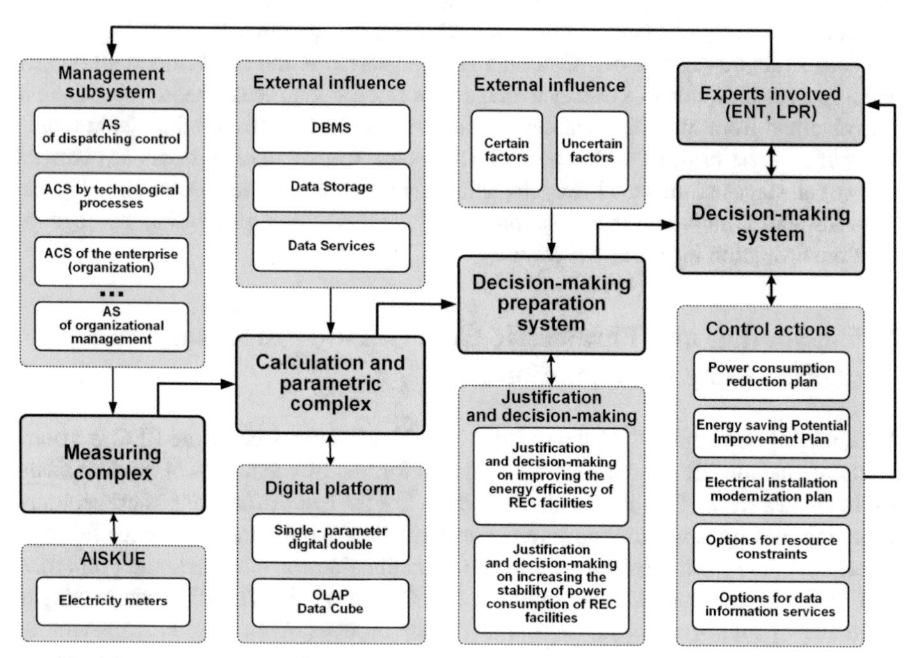

Fig. 1. The structure of the process reflecting the various states of data on power consumption within the framework of the energy management process.

Then, within the digital profile, with the help of software functionality based on vector rank analysis, the parameter data is processed in the RPC. The purpose of working with data is to obtain deterministic invariants that enter the analytical subsystem and then with materials into the decision support system (DSS), which develops proposals to the management subsystem for the formation of control actions. In this article, the software functionality of the RPC based on vector rank analysis will be considered in detail.

The functioning of the elements of the RPC is based on the constant exchange of data with the database of data on power consumption, the main element of which is the data warehouse for the studied parameter (display, process). The storage of data on power consumption is understood as a subject-oriented information database of the REC used to support the adoption of solutions within the framework of the digital energy efficiency platform [5]. Its main structural unit is a digital data layer - a flat data structure (a two-dimensional array) that stores a set of parameters identified by index and/or rank (the first dimension), as well as the number of the time interval (the second dimension), which are (as a rule) the result of cyclic implementation of software functionality on an OLAP cube data on power consumption [5]. The digital data layer acts as the main horizontal structural unit of the OLAP cube.

Figure 2 shows the composition of the digital layers of the REC data warehouse on electricity consumption, which are formed during the implementation of methods, models and techniques of the digital energy efficiency platform.

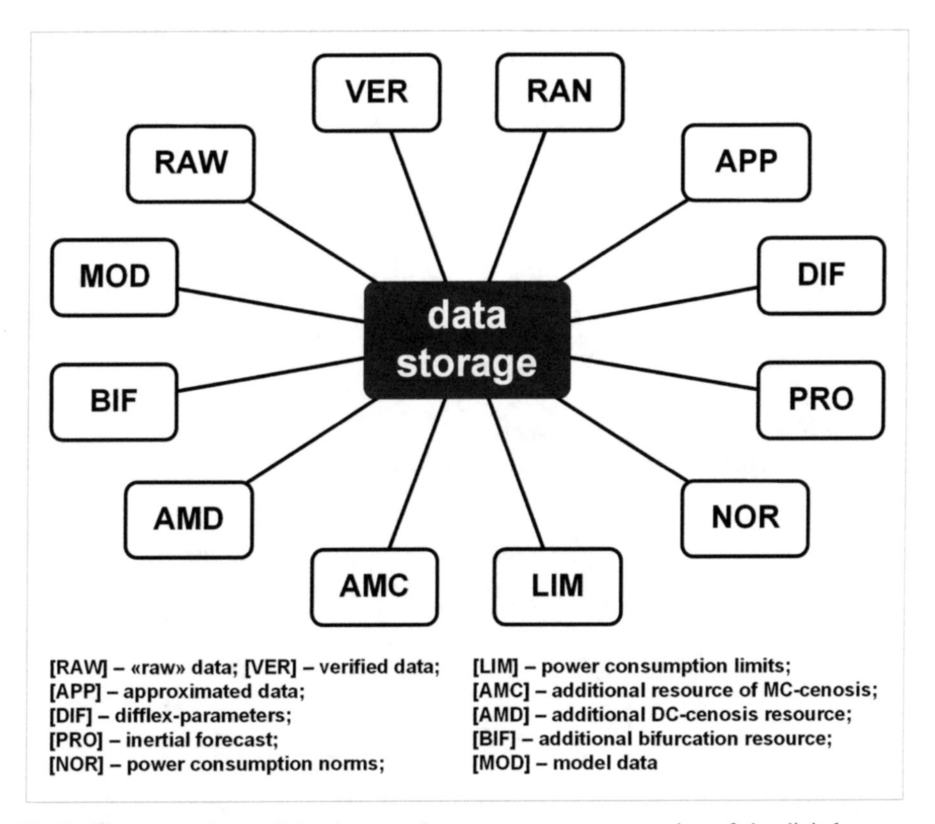

Fig. 2. The composition of the data warehouse on power consumption of the digital energy efficiency platform.

Digital layers include "raw" data – [RAW]; verified data – [VER]; ranked data – [RAN]; approximated data – [APP]; difflex parameters - [DIF]; forecast data – [PRO]; [NOR] – power consumption standards – [NOR]; electrical consumption limits - [LIM]; synthesized data – [SIN], resulting from the synthesis of rank analysis procedures; model data as a result of taking into account the control effect in static modeling based on vector rank analysis – [MOD]; the additional resource of MC-cenosis obtained within the framework of the vector bifurcation model is [AMC]; the additional resource of Z2–energy saving potential as a result of the vector bifurcation model is [POT].

It should be noted that the digital layers [DIF], [PRO], [NOR] and [LIM] are formed according to the results of standard procedures: interval estimation, forecasting, normalization and hashing. In addition, the primary layers have the same structure as the data warehouse when implementing functional rank analysis methods [3]. The difference is made by the layers obtained as a result of the implementation of vector rank analysis methods: [SIN], [MOD], [AMC] and [POT].

The first four layers (RAW, VER, RAN, APP) are formed as a result of collecting, formatting, cleaning and parametric adaptation of power consumption data [8]. Digital layers [SIN] and [MOD] are formed during the implementation of the method of controlling the electrical consumption of REC facilities based on the synthesis of rank analysis procedures and are the result of static modeling, as well as taking into account the internal control action. The implementation of the power consumption management method of REC facilities based on an additional resource replenishes the data storage with digital layers [AMC] and [POT].

The mathematical description of the digital data layer obtained as a result of vector rank analysis looks as follows:

$$\left\langle W_{kt}^{OLAP} \right\rangle \xrightarrow[\substack{k=1..n \\ t=1..\tau}]{p=fix} \left\langle \begin{matrix} \overset{kt}{[VER]_{kt}} & \overset{kt}{[PRO]_{kt}} & \overset{kt}{[MOD]_{kt}} \\ [RAN]_{kt} & [NOR]_{kt} & [AMC]_{kt} \\ [APP]_{kt} & [LIM]_{kt} & [POT]_{kt} \end{matrix} \right\rangle, \tag{1}$$

$\left\langle W_{kt}^{OLAP} \right\rangle$ – the OLAP data cube tuple

Digital layers can be represented as a two-dimensional array:

$$\left[W_{kt}^{VER} \right] \xrightarrow[t=1..\tau]{k=1..n} \begin{bmatrix} W_{11}^{VER} & W_{12}^{VER} & W_{13}^{VER} & \ldots & W_{1j}^{VER} & \ldots & W_{1\tau}^{VER} \\ W_{21}^{VER} & W_{22}^{VER} & W_{23}^{VER} & \ldots & W_{2j}^{VER} & \ldots & W_{2\tau}^{VER} \\ W_{31}^{VER} & W_{32}^{VER} & W_{33}^{VER} & \ldots & W_{3j}^{VER} & \ldots & W_{3\tau}^{VER} \\ \ldots & \ldots & \ldots & \ldots\ldots & & \ldots\ldots & \\ W_{i1}^{VER} & W_{i2}^{VER} & W_{i3}^{VER} & \ldots & W_{ij}^{VER} & \ldots & W_{i\tau}^{VER} \\ \ldots & \ldots & \ldots & \ldots\ldots & & \ldots\ldots & \\ W_{n1}^{RAN} & W_{n2}^{RAN} & W_{n3}^{RAN} & \ldots & W_{nj}^{RAN} & \ldots & W_{n\tau}^{RAN} \end{bmatrix}, \tag{2}$$

$\left[W_{kt}^{VER} \right]$ – two-dimensional matrix of the digital layer of verified data on power consumption (example); W_{kt}^{p} – power consumption; p – id of the data layer; k – rank (n is the total number of REC objects); t – time interval (τ – total number of model time intervals).

For the purpose of interactive or operational analysis of the energy efficiency of REC facilities, an OLAP cube is formed on the basis of the data warehouse, a multi-dimensional, long-term stored array of REC data on power consumption, used in the process of interactive energy efficiency analysis (Fig. 3) [5].

Mathematically, the representation of the OLAP cube of REC data on power consumption can be described using the following system:

$$\left\langle W_{kt}^{OLAP} \right\rangle \xrightarrow[t=1..\tau]{k=1..n} \left\langle \begin{matrix} [RAW]_{kt} & [DIF]_{kt} & [SIN]_{kt} & [IPK]_{kt} \\ [VER]_{kt} & [PRO]_{kt} & [MOD]_{kt} & [IPZ]_{kt} \\ [RAN]_{kt} & [NOR]_{kt} & [AMC]_{kt} & [DFU]_{kt} \\ [APP]_{kt} & [LIM]_{kt} & [POT]_{kt} & [IPE]_{kt} \end{matrix} \right\rangle; \tag{3}$$

Rank	Data of the REC OLAP cube layer on power consumption										
	...	1	2	3	4	5	6	......	t	......	τ
1	...	W_{11}	W_{12}	W_{13}	W_{14}	W_{15}	W_{16}	......	W_{1t}	......	$W_{1\tau}$
2	...	W_{21}	W_{22}	W_{23}	W_{24}	W_{25}	W_{26}	......	W_{2t}	......	$W_{2\tau}$
3	...	W_{31}	W_{32}	W_{33}	W_{34}	W_{35}	W_{36}	......	W_{3t}	......	$W_{3\tau}$
4	...	W_{41}	W_{42}	W_{43}	W_{44}	W_{45}	W_{46}	......	W_{4t}	......	$W_{4\tau}$
5	...	W_{51}	W_{52}	W_{53}	W_{54}	W_{55}	W_{56}	......	W_{5t}	......	$W_{5\tau}$
......	...	......	......	......	......	......	......	......	......	......	......
k	...	W_{k1}	W_{k2}	W_{k3}	W_{k4}	W_{k5}	W_{k6}	......	W_{kt}	......	$W_{k\tau}$
......	...	......	......	......	......	......	......	......	......	......	......
n-1	...	$W_{(n-1)1}$	$W_{(n-1)2}$	$W_{(n-1)3}$	$W_{(n-1)4}$	$W_{(n-1)5}$	$W_{(n-1)6}$	......	$W_{(n-1)t}$	......	$W_{(n-1)\tau}$
n	...	W_{n1}	W_{n2}	W_{n3}	W_{n4}	W_{n5}	W_{n6}	......	W_{nt}	......	$W_{n\tau}$

Fig. 3. OLAP-REC data cube by power consumption parameter.

where are the primary aggregators :

$$\left\{\begin{array}{l} w:\{[RAW],[VER],[RAN]\} \rightarrow [APP]; \\ w:\{[VER],[RAN],[APP]\} \rightarrow [DIF]; \\ w:\{[VER],[RAN],[APP]\} \rightarrow [PRO]; \\ w:\{[VER],[RAN],[APP]\} \rightarrow [NOR]; \\ w:\{[VER],[DIF],[PRO]\} \rightarrow [LIM]; \\ w:\{[APP],[DIF],[PRO]\} \rightarrow [AMC]; \end{array}\right.$$

secondary aggregators :

$$\left\{\begin{array}{l} w:\{[APP],[DIF],[PRO]\} \rightarrow [POT]; \\ w:\{[APP],[DIF],[POT]\} \rightarrow [IPK]; \\ w:\{[APP],[DIF],[POT]\} \rightarrow [IPZ]; \\ w:\{[APP],[IPK],[IPZ]\} \rightarrow [IPE]; \\ w:\{[APP],[DIF],[IPE]\} \rightarrow [DFU]; \\ w:\{[APP],[DIF],[IPE]\} \rightarrow [DAM], \end{array}\right.$$

$\langle W_{kt}^{OLAP} \rangle$ – the OLAP data cube tuple; $w:\{[S1],[S2]\} \rightarrow [S3]$ – designation of the aggregation operation of two primary layers [S1] and [S2] to the secondary [S3].

Expression (3) presents these aggregators. In addition, new secondary layers are introduced in this expression: [IPK] – integral quality indicators; [IPZ] – integral cost indicators; [IPE] – integral efficiency indicators and [DFU] – difflex angles of objects.

In the process of forming an OLAP cube of REC data according to the parameter of electricity consumption, data cubing is carried out, which is a certain way of creating, long-term storage, as well as interactive analysis of a multidimensional array of REC data on electricity consumption in the form of an OLAP cube. Figure 4 schematically shows the elements of an OLAP cube.

By examining and analyzing the OLAP cube of REC data by the parameter of power consumption, it is possible to determine new structural elements in it. These include: a digital double of the power consumption of the REC, a digital vector of the rank (object), a digital slice of the REC, a digital double of the rank of the REC, a digital double of the object of the REC.

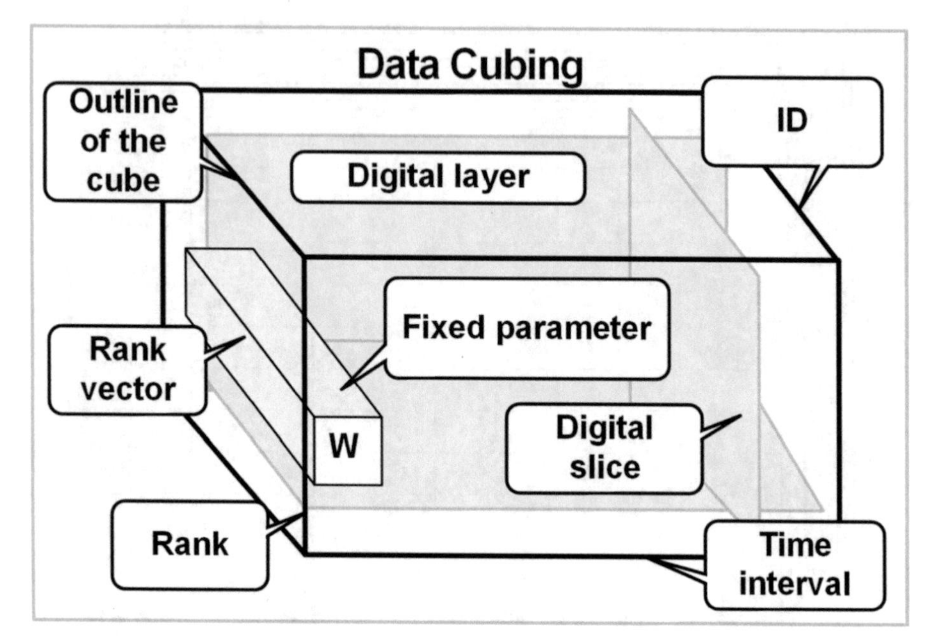

Fig. 4. Elements of the OLAP data cube.

The digital double of power consumption of the REC is a digital profile that is constantly changing under the influence of software functionality, containing an actual storage of digital doubles of power consumption, cubed into a set of digital rank vectors [5]. At the same time, the software functionality of the REC is understood as a tuple of converters and aggregators designed for the implementation of static, dynamic and bifurcation models, as well as the procedures for the reverse adaptation on the OLAP cube of power consumption data.

The mathematical record of the elements of the digital double in terms of electrical consumption is presented in expression (4) and contains tuples of transactions of ranks (objects) in terms of power consumption, as well as digital vectors of ranks (objects) in terms of power consumption of the REC.

Expression (4) includes groups of layers of the digital twin: PO –primary data analysis, SM – static modeling (interval estimation, difflex analysis, forecasting, GZ analysis, normalization, ASR analysis), SP – synthesis of rank analysis procedures (monitoring, aggregation, accounting for the control effect), DR – additional resources (MC-forecasting, assessment of Z2-energy saving potential), RN – regime rationing (construction of R3,-R2,-R3-distributions, establishment of R3,-R2,-R3-modes), KL – planning of energy saving measures (hashing, ZP analysis).

$$\begin{cases} \langle W_{1t}^{OLAP} \rangle \underset{k=1;\ t=1..\tau}{\longrightarrow} \langle \langle W_{1t}^{PO} \rangle,\ \langle W_{1t}^{SM} \rangle,\ \langle W_{1t}^{SP} \rangle,\ \langle W_{1t}^{DR} \rangle,\ \langle W_{1t}^{RN} \rangle,\ \langle W_{1t}^{KL} \rangle \rangle; \\ \langle W_{2t}^{OLAP} \rangle \underset{k=2;\ t=1..\tau}{\longrightarrow} \langle \langle W_{2t}^{PO} \rangle,\ \langle W_{2t}^{SM} \rangle,\ \langle W_{2t}^{SP} \rangle,\ \langle W_{2t}^{DR} \rangle,\ \langle W_{2t}^{RN} \rangle,\ \langle W_{2t}^{KL} \rangle \rangle; \\ \langle W_{3t}^{OLAP} \rangle \underset{k=3;\ t=1..\tau}{\longrightarrow} \langle \langle W_{3t}^{PO} \rangle,\ \langle W_{3t}^{SM} \rangle,\ \langle W_{3t}^{SP} \rangle,\ \langle W_{3t}^{DR} \rangle,\ \langle W_{3t}^{RN} \rangle,\ \langle W_{3t}^{KL} \rangle \rangle; \\ \ldots\ \ldots\ \ldots \\ \left\langle W_{jt}^{OLAP} \right\rangle \underset{k=j;\ t=1..\tau}{\longrightarrow} \left\langle \left\langle W_{jt}^{PO} \right\rangle,\ \left\langle W_{jt}^{SM} \right\rangle,\ \left\langle W_{jt}^{SP} \right\rangle,\ \left\langle W_{jt}^{DR} \right\rangle,\ \left\langle W_{jt}^{RN} \right\rangle,\ \left\langle W_{jt}^{KL} \right\rangle \right\rangle; \\ \ldots\ \ldots\ \ldots \\ \langle W_{nt}^{OLAP} \rangle \underset{k=n;\ t=1..\tau}{\longrightarrow} \langle \langle W_{nt}^{PO} \rangle,\ \langle W_{nt}^{SM} \rangle,\ \langle W_{nt}^{SP} \rangle,\ \langle W_{nt}^{DR} \rangle,\ \langle W_{nt}^{RN} \rangle,\ \langle W_{nt}^{KL} \rangle \rangle; \end{cases} \tag{4}$$

$$\left\langle W_{t}^{OLAP} \right\rangle \underset{t=1..\tau}{\longrightarrow} \left\langle \begin{matrix} \left\langle W_{1t}^{VEC} \right\rangle,\ \left\langle W_{2t}^{VEC} \right\rangle, \\ \left\langle W_{3t}^{VEC} \right\rangle,\ \ldots,\ \left\langle W_{jt}^{VEC} \right\rangle \\ ,\ \ldots,\ \left\langle W_{nt}^{VEC} \right\rangle \end{matrix} \right\rangle,$$

$\langle W_{t}^{OLAP} \rangle$– the tuple of the OLAP cube of REC data on the t-th time interval; $\langle W_{kt}^{OLAP} \rangle$ – the tuple of a k-th rank transaction on the t-th time interval; $\langle W_{kt}^{s} \rangle$ – the tuple of the s-th group of layers of the k-th rank cube on the t-th time interval (PO, SM, SP, DR, RN, KL); $\langle W_{kt}^{VEC} \rangle$ – a digital vector of the k-th rank on the t-th time interval

The digital vector of the rank (object) is a tuple of transactions containing up–to-date adapted data on the power consumption of the rank (object), which are the result of the cyclical implementation of the software functionality of technocenosis on an OLAP data cube [5]. It allows you to reflect on the required time interval all the information about the power consumption of the rank (object) of the REC. In addition, the digital vector is the main vertical structural unit of the OLAP cube. Mathematically, in the form of a tuple of transactions of the k-th rank of the REC on the t-th time interval, it can be described as follows:

$$\left\langle W_{kt}^{OLAP} \right\rangle \underset{k=fix;\ t=fix}{\longrightarrow} \left\langle \begin{matrix} \left\langle W_{kt}^{PO} \right\rangle,\ \left\langle W_{kt}^{SM} \right\rangle, \\ \left\langle W_{kt}^{SP} \right\rangle,\ \left\langle W_{kt}^{DR} \right\rangle, \\ \left\langle W_{kt}^{RN} \right\rangle,\ \left\langle W_{kt}^{KL} \right\rangle \end{matrix} \right\rangle; \tag{5}$$

$\langle W_{kt}^{OLAP} \rangle$ – the tuple of a k-th rank transaction on the t-th time interval.

The digital slice of the REC is a flat data structure (usually a two–dimensional array) that stores a set of parameters identified by index and/or rank (the first dimension), as well as the identifier of the data layer (the second dimension), which are the result of cyclic implementation of the software functionality on the OLAP data cube (6) [5]. It allows you to fully display information about all ranks of the VRP. Digital cross-section of technocenosis data on the parameter of power consumption:

$$\left\langle W_{kt}^{OLAP} \right\rangle \xrightarrow[t=1..\tau]{k=1..n} \left\langle \begin{matrix} [RAW]_{kt} & [DIF]_{kt} & [SIN]_{kt} & [IPK]_{kt} \\ [VER]_{kt} & [PRO]_{kt} & [MOD]_{kt} & [IPZ]_{kt} \\ [RAN]_{kt} & [NOR]_{kt} & [AMC]_{kt} & [DFU]_{kt} \\ [APP]_{kt} & [LIM]_{kt} & [POT]_{kt} & [IPE]_{kt} \end{matrix} \right\rangle, \tag{6}$$

and a two-dimensional array of digital data slice on power consumption can be represented as follows:

$$\left[W_{kt}^{p} \right] \xrightarrow[p=1..m]{k=1..n,\ t=T} \begin{bmatrix} W_{1T}^{1} & W_{1T}^{2} & W_{1T}^{3} & \cdots & W_{1T}^{j} & \cdots & W_{1T}^{m-1} & W_{1T}^{m} \\ W_{2T}^{1} & W_{2T}^{2} & W_{2T}^{3} & \cdots & W_{2T}^{j} & \cdots & W_{2T}^{m-1} & W_{2T}^{m} \\ W_{3T}^{1} & W_{3T}^{2} & W_{3T}^{3} & \cdots & W_{3T}^{j} & \cdots & W_{3T}^{m-1} & W_{3T}^{m} \\ \cdots & \cdots & \cdots & \cdots\cdots & & \cdots\cdots & & \cdots \\ W_{iT}^{1} & W_{iT}^{2} & W_{iT}^{3} & \cdots & W_{iT}^{j} & \cdots & W_{iT}^{m-1} & W_{iT}^{m} \\ \cdots & \cdots & \cdots & \cdots\cdots & & \cdots\cdots & & \cdots \\ W_{nT}^{1} & W_{nT}^{2} & W_{nT}^{3} & \cdots & W_{nT}^{j} & \cdots & W_{nT}^{m-1} & W_{nT}^{m} \end{bmatrix},$$

$\left\langle W_{kp}^{OLAP} \right\rangle$ – OLAP-a cube of data on power consumption at a time t = T;

$\left[W_{kt}^{p} \right]$ – the matrix of the digital slice of the OLAP data cube; W_{kT}^{p} – the value of the parameter corresponding to the k-th rank at time t = T in the p-th data layer of the cube; k – rank; p – identifier of the OLAP cube layer.

The digital double of the REC rank is a flat data structure (usually a two–dimensional array) that stores a set of parameters identified by the identifier of the data layer (the first dimension), as well as the number of the time interval (the second dimension), which are the result of the cyclic implementation of the software functionality on the OLAP cube of power consumption data in relation to the selected rank (7) [5]. It allows you to fully parametrically describe the allocated rank of the REC during the entire time. Digital double of the rank of the REC according to the parameter of power consumption:

$$\left\langle W_{kt}^{OLAP} \right\rangle \xrightarrow[t=1..\tau]{k=R} \left\langle \begin{matrix} [RAW]_{Rt} & [DIF]_{Rt} & [SIN]_{Rt} & [IPK]_{Rt} \\ [VER]_{Rt} & [PRO]_{Rt} & [MOD]_{Rt} & [IPZ]_{Rt} \\ [RAN]_{Rt} & [NOR]_{Rt} & [AMC]_{Rt} & [DFU]_{Rt} \\ [APP]_{Rt} & [LIM]_{Rt} & [POT]_{Rt} & [IPE]_{Rt} \end{matrix} \right\rangle; \tag{7}$$

a two-dimensional array of a digital double of rank in terms of power consumption:

$$\left[W_{kt}^{p} \right] \xrightarrow[p=1..m]{k=R,\ t=1..\tau} \begin{bmatrix} W_{R1}^{1} & W_{R1}^{2} & W_{R1}^{3} & \cdots & W_{R1}^{j} & \cdots & W_{R1}^{m-1} & W_{R1}^{m} \\ W_{R2}^{1} & W_{R2}^{2} & W_{R2}^{3} & \cdots & W_{R2}^{j} & \cdots & W_{R2}^{m-1} & W_{R2}^{m} \\ W_{R3}^{1} & W_{R3}^{2} & W_{R3}^{3} & \cdots & W_{R3}^{j} & \cdots & W_{R3}^{m-1} & W_{R3}^{m} \\ \cdots & \cdots & \cdots & \cdots\cdots & & \cdots\cdots & & \cdots \\ W_{Ri}^{1} & W_{Ri}^{2} & W_{Ri}^{3} & \cdots & W_{Ri}^{j} & \cdots & W_{Ri}^{m-1} & W_{Ri}^{m} \\ \cdots & \cdots & \cdots & \cdots\cdots & & \cdots\cdots & & \cdots \\ W_{R\tau}^{1} & W_{R\tau}^{2} & W_{R\tau}^{3} & \cdots & W_{R\tau}^{j} & \cdots & W_{R\tau}^{m-1} & W_{R\tau}^{m} \end{bmatrix},$$

$\langle W_{kt}^{OLAP} \rangle$ – OLAP-data cube for the rank value k = R; $[W_{kt}^{p}]$ – matrix of a digital double of rank k = R OLAP-data cube; W_{Rt}^{p} – the value of the parameter corresponding to the p-th data layer for rank k = R at the t-th moment of time; k is the rank; t is the time interval (– total number of intervals); p is the identifier of the OLAP cube data layer (m is the total number of data layers).

The digital twin of the REC object is a flat data structure (as a rule, a two-dimensional array) that stores a set of parameters identified by the identifier of the data layer (the first dimension), as well as the number of the time interval (the second dimension), which are the result of the cyclic implementation of the software functionality on the OLAP cube of power consumption data in relation to the selected object (8) [5]. The digital double of the REC object allows you to fully describe the parameters of the object throughout the entire time interval. Digital double of the REC facility for power consumption:

$$\left\langle W_{qt}^{OLAP} \right\rangle \xrightarrow[t=1..\tau]{q=B} \left\langle \begin{matrix} [RAW]_{Bt} & [DIF]_{Bt} & [SIN]_{Bt} & [IPK]_{Bt} \\ [VER]_{Bt} & [PRO]_{Bt} & [MOD]_{Bt} & [IPZ]_{Bt} \\ [RAN]_{Bt} & [NOR]_{Bt} & [AMC]_{Bt} & [DFU]_{Bt} \\ [APP]_{Bt} & [LIM]_{Bt} & [POT]_{Bt} & [IPE]_{Bt} \end{matrix} \right\rangle, \tag{8}$$

and a two - dimensional array of a digital double of an object in terms of power consumption can be represented as follows:

$$\left[W_{qt}^{p}\right] \xrightarrow[p=1..m]{q=B,\ t=1..\tau} \begin{bmatrix} W_{B1}^{1} & W_{B1}^{2} & W_{B1}^{3} & \cdots & W_{B1}^{j} & \cdots & W_{B1}^{m-1} & W_{B1}^{m} \\ W_{B2}^{1} & W_{B2}^{2} & W_{B2}^{3} & \cdots & W_{B2}^{j} & \cdots & W_{B2}^{m-1} & W_{B2}^{m} \\ W_{B3}^{1} & W_{B3}^{2} & W_{B3}^{3} & \cdots & W_{B3}^{j} & \cdots & W_{B3}^{m-1} & W_{B3}^{m} \\ \cdots & \cdots & \cdots & \cdots\cdots & & \cdots\cdots & & \cdots \\ W_{Bi}^{1} & W_{Bi}^{2} & W_{Bi}^{3} & \cdots & W_{Bi}^{j} & \cdots & W_{Bi}^{m-1} & W_{Bi}^{m} \\ \cdots & \cdots & \cdots & \cdots\cdots & & \cdots\cdots & & \cdots \\ W_{B\tau}^{1} & W_{B\tau}^{2} & W_{B\tau}^{3} & \cdots & W_{B\tau}^{j} & \cdots & W_{B\tau}^{m-1} & W_{B\tau}^{m} \end{bmatrix},$$

$\langle W_{qt}^{OLAP} \rangle$ – OLAP-a data cube for an object that has an index q = B; $[W_{qt}^{p}]$ – the matrix of the digital double of an object having an index q = B; W_{Bt}^{p} – the value of the parameter corresponding to the p-th data layer for an object with the index q = B; q is the index of the object (n is the total number of objects); t is the time interval (τ – total number of intervals); p is the ID of the OLAP cube data layer (m is the total number of data layers).

Thus, the concept of parametric virtualization is introduced on the basis of the concept of data storage, as well as the definitions of the digital double of the power consumption of the REC, the digital vector of the rank (object), the digital slice of the REC, the digital double of the rank of the REC, the digital double of the REC object. It is a way of creating a digital twin of RIVERS, involving the formation and processing of a data warehouse using rank analysis procedures.

3 Algorithmic Structure of the Digital Energy Efficiency Platform

It is shown in [3] that electric heating is a purposeful and controlled process implemented in order to provide electricity in the required quantity and required quality with the maximum possible energy savings and minimization of expenses. The proposed parametric digital twin for the first time makes it possible to significantly improve the quality of REC management in conditions of limited capabilities of the regional generating complex.

Currently, digital transformation strategies are being rapidly implemented at electric power facilities. Their content suggests that the transition to digital energy will significantly improve the quality of energy resource management in conditions of limited capabilities of regional generating complexes. For this purpose, a number of strategies, concepts and roadmaps have been developed and implemented. One of the most promising is the concept of the Internet of Energy, developed by a team of authors led by D.A. Kholkin, I.S. Chausov and I.A. Burdin. The concept of the Internet of Energy is a type of decentralized electric power system that implements intelligent distributed management based on transactions between its users. It is based on the principle of interaction between 4D energy cells: Decarbonization (decarbonization), Decentralization (decentralization), Diversification (diversification). In [9], the authors describe in detail all the elements of this concept. However, the authors pay insufficient attention to the element, which involves the implementation of software applications for the implementation of energy transactions between users of energy cells. The use of a parametric digital twin in the management of energy transactions between the users of energy cells will make it possible to implement a fully digital energy efficiency board in various operating conditions of the REC.

Figure 5 shows the algorithmic structure of the digital energy efficiency platform based on vector rank analysis [10].

Its implementation begins with the import of power consumption data into the data warehouse after their processing in the measuring complex (Fig. 1). It is shown above that the main structural unit of the data warehouse is the digital layer. The formation of digital layers begins with parametric adaptation of power consumption data, as a result of which they are cleaned, formatted, verified and approximated, as well as the choice of a reference layer for power consumption. Based on the selected reference layer, a static model of functional rank analysis is implemented, which assumes obtaining vectors of primary procedures for interval estimation, forecasting, rationing and potency of electricity consumption. Based on the obtained vectors, which are transformed using interactive data analysis into a data warehouse, an OLAP cube of data on the power consumption of the REC is created. Further, power consumption management methods based on vector rank analysis are implemented [10]. After loading the results of the procedures of interval estimation, forecasting and rationing of power consumption, they are monitored for errors and deviations. In case of detection, their elimination is implemented using complex algorithms. Upon completion of the elimination of errors and deviations, a refined C-matrix is formed. At the next stage, a vector entropy model is implemented, within which, based on the selected refinement procedure, additional resources are calculated in order to increase the efficiency of power consumption management.

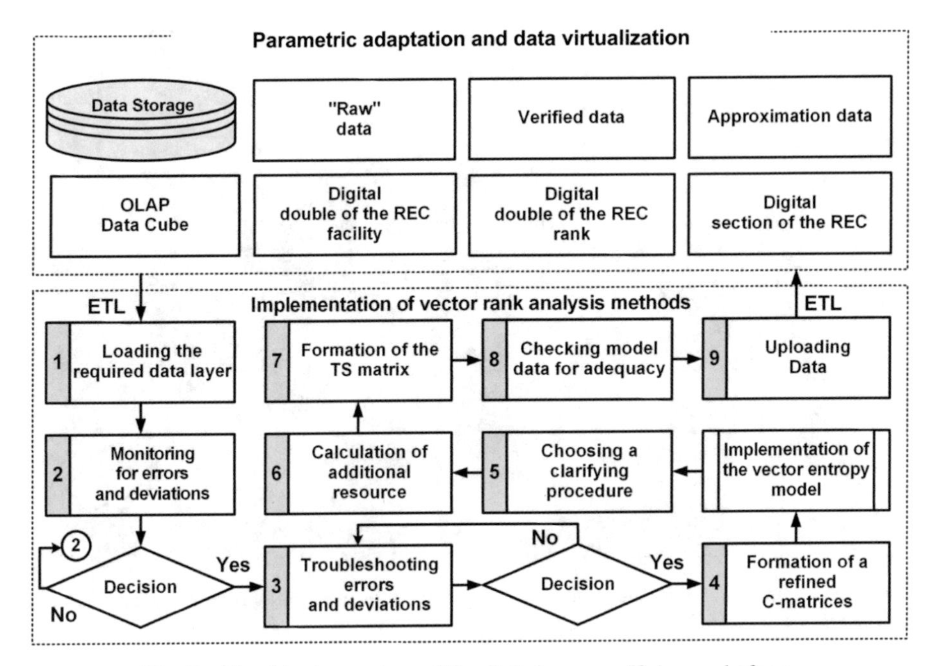

Fig. 5. Algorithmic structure of the digital energy efficiency platform.

Thus, at the final stage of the algorithm for implementing the digital energy efficiency board, the results obtained are checked for adequacy using integral criteria of predictive capabilities and deterministic invariants for the DSS are formed.

The implementation of the presented algorithm was carried out in the software and hardware complex (PAK) for monitoring the power consumption of the REC of JSC Yan-Tarenergo [11]. The complex is a system of automated workstations (APMs) developed in C# and a geoinformation system for energy facilities in the Kaliningrad region (Fig. 6).

The automated control systems include algorithms of the digital energy efficiency platform for performing procedures for interval estimation, forecasting, rationing, potency, constructing typical graphs of electrical load and trends, as well as re-rationing. Figures 7, 8 and 9 show the fragments of the automated control systems implementing the algorithms of the digital energy efficiency platform.

The simplest and most convenient use of all the software functionality of the digital platform by PAK operators is carried out using an OLAP cube, which allows real-time processing of a multidimensional, long-term stored array of data on the power consumption of the REC.

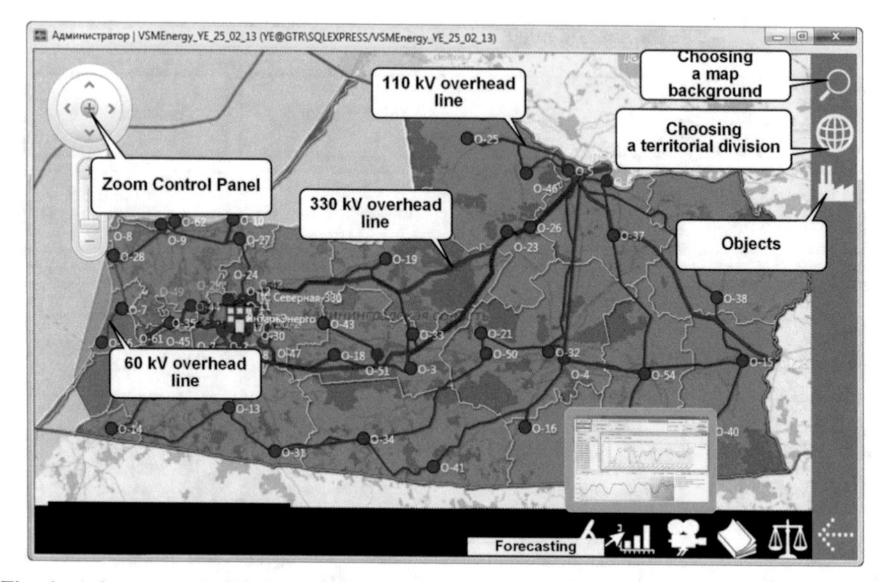

Fig. 6. A fragment of a hardware and software complex for monitoring power consumption of a regional electrical complex (geo-information system).

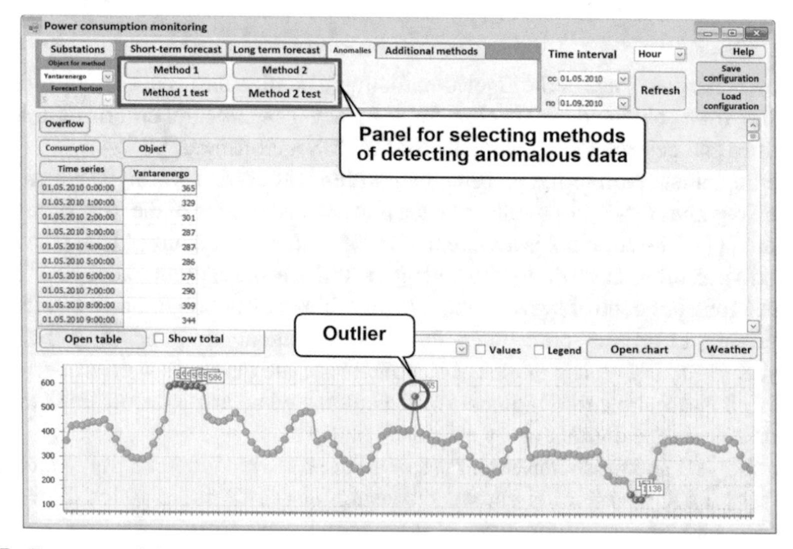

Fig. 7. Fragment of the automated workplace of the interval evaluation procedure forming the [DIF] layer.

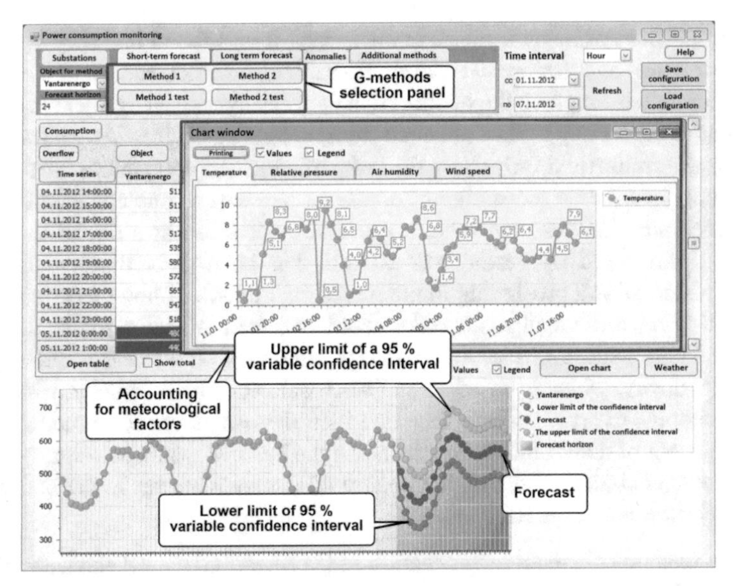

Fig. 8. A fragment of the automated workplace of the forecasting procedure forming the [PRO] layer.

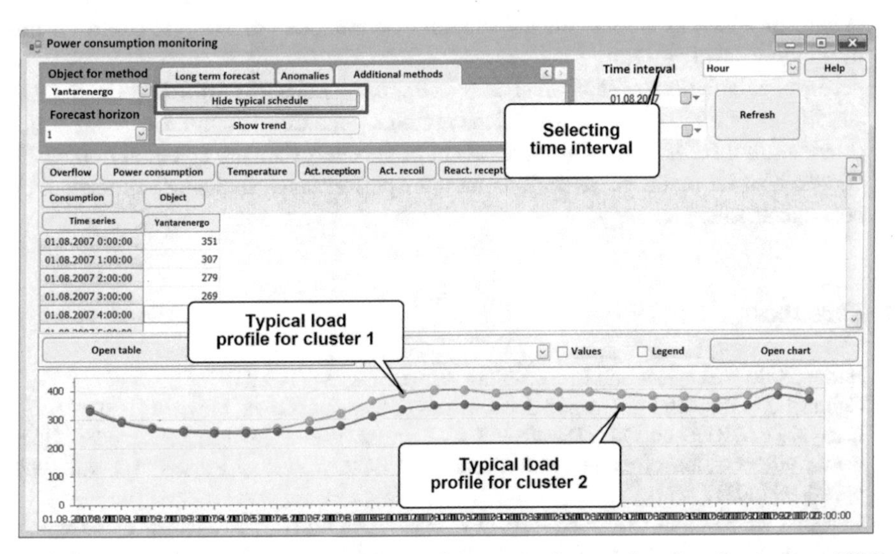

Fig. 9. Fragment of the automated workplace of the synthesis procedure forming the layer [SIN].

4 Conclusion

Thus, the main purpose of energy resource management of the REC is to ensure sustainable power consumption in various modes of its operation. As a result of the installation of a large number of digital devices at energy facilities, it became possible to record significant volumes of measurements, which, in the future, can be converted into data.

Based on the study of many samples of such data, it was found that they have a non-standard structure and are non-Gaussian. Therefore, the most important element that makes it possible to improve the quality of energy resources management is the scientifically based processing of data obtained from measuring devices in real time in order to develop deterministic invariants on the basis of which control actions are formed.

Currently, qualitative processing of non-Gaussian data is implemented using rank analysis methods. For the first time, a procedure for creating a digital energy efficiency platform based on vector rank analysis has been theoretically substantiated for REC, which allows processing multidimensional arrays of non-Gaussian data and implementing scientific methods in order to obtain deterministic invariants for DSS.

A distinctive feature of the digital energy efficiency platform is a one-parameter digital REC binary, which is based on an OLAP data cube by parameter, which allows long-term storage of arrays of non-Gaussian REC data on power consumption used for the development of deterministic invariants, on the basis of which decisions are made.

The implementation of the digital energy efficiency platform in the software and hardware complexes of the REC will allow:

- perform cleaning, formatting, processing and verification of power consumption data received via various communication channels from automated metering devices;
- to implement in real time continuous accounting and monitoring of data on power consumption, a comprehensive assessment of the impact of REC on objects and objects on REC, as well as the calculation of parameters of external control action by a higher-level system;
- perform calculation of technical and economic indicators of REC facilities in conditions of limited capabilities of the regional generating complex;
- provide full information on the assessment of the energy saving potential of the REC and each of its facilities, as well as monitor performance indicators and select the best management strategy.

References

1. Haitun, S.D.: Mechanics and Irreversibility. Janus, Moscow (1996)
2. Haitun, S.D.: Problems of Quantitative Analysis of Science. Nauka, Moscow (1989)
3. Gnatyuk, V.I., Kivchun, O.R., Dorofeev, S.A., Bovtrikova, E.V.: Mathematical model of parametric virtualization of technocenosis data. In: CEUR Workshop Proceedings, vol. 2922, pp. 90–99 (2021)
4. Gnatyuk, V.I. Kivchun, O.R., Lutsenko, D.V.: Digital platform for management of the regional power grid consumption. In: IOP Conference Series: Earth and Environmental Science, vol. 689, no. 1, p. 012022 (2021)
5. Kholkin, D.V.: People Figures. Six Views on New Energy. Litagent Ridero, Moscow (2020)

Development of a Probabilistic Model for Predicting Advertising Conversion Using Recurrent Neural Networks

V. V. Savenkov[1], E. O. Bobrova[1](✉), E. N. Shcherbak[2], A. E. Shcherbak[3], I. V. Tarasov[4], and L. V. Dimitrov[5]

[1] Moscow Polytechnic University, 38 st. Bolshaya Semyonovskaya, Moscow 107023, Russia
saaturn2015@mail.ru

[2] Russian State University for the Humanities (RSUH), 6 Miusskaya sq., Moscow 125993, Russia

[3] First Data LLC, 30 st. Pokrovka, Moscow 105062, Russia

[4] State Budget Educational Institution School Number 1288 Named After the Hero of the Soviet Union N.V., Troyan, 3a Polykarpova Street, Moscow 125284, Russia

[5] Technical University of Sofia, 25 Tsanko Diustabanov St., Plovdiv 4000, Bulgaria

Abstract. This article covers a comprehensive study in the field of website traffic prediction where machine learning is used. It also additionally focuses on a specialized neural network structure utilizing LSTM layers and an attention layer, explicitly crafted for the processing of time series data. The paper extensively discusses LSTM layers and their ability to capture long-term dependencies in data, as well as the integration of an attention layer to focus on key moments. The article also includes the stages of model training, from hyperparameter tuning to evaluating results on test data. Hyperparameters such as hidden layer size, initial learning rate, and others were carefully tuned to strike a balance between performance and training resources. Model evaluation was conducted using mean absolute error (MAE) and mean squared error (MSE).

Keywords: Online Advertising · Probabilistic Models · Web Analytics · Campaign Optimization · Data Analysis · Advertising Efficiency · Conversions · A/B Testing

1 Introduction

To achieve optimal results, it is necessary to use advanced methods of data analysis and modeling. The use of probabilistic models and web analytics is proposed to optimize online advertising.

The project's goal is to enhance the efficiency of online advertising by optimizing advertising campaigns. To attain this objective, we will employ probabilistic models grounded in statistical methods and conduct data analysis using web analytics [3]. These probabilistic models enable the incorporation of diverse factors, including target audience, context, and user preferences, enhancing the precision of predicting advertising

A. Gibadullin (Ed.): DITEM 2023, LNNS 942, pp. 233–242, 2024.
https://doi.org/10.1007/978-3-031-55349-3_19

campaign outcomes. Data analysis collected through web analytics will provide valuable information about user behavior, the effectiveness of advertising channels, conversions, and the impact of various factors on advertising success [11].

The primary objectives of the project involve gathering and assessing data related to user behavior, advertising channels, conversions, and various metrics associated with advertising effectiveness. Detailed data analysis will be conducted using web analytics methods to identify key factors influencing the success of advertising campaigns. Subsequently, predicting advertising campaign results will involve developing probabilistic models using the provided data. These models will consider factors like geographic location, demographic characteristics, user interests, and preferences.

Optimizing advertising strategies will involve utilizing these models, which includes optimizing the allocation of the budget, selecting the most effective advertising channels, and crafting personalized advertising messages [9]. A/B testing of various advertising campaign variations will also be conducted to determine the most successful approaches. Afterward, optimized advertising strategies will be tested and evaluated based on key metrics such as conversions, return on investment (ROI), and profit.

Finally, the project involves continuous monitoring and analysis of advertising campaign results, as well as making adjustments to strategies to achieve maximum efficiency. This includes regular updates and analysis of data, monitoring competitors and changes in the market environment, and applying machine learning and optimization algorithms to automate decision-making processes.

In conclusion, this project on optimizing online advertising through probabilistic models and web analytics will increase conversions, maximize profits, and improve the targeting of advertising campaigns. This will be achieved through the use of advanced methods of data analysis and modeling, as well as continuous monitoring and optimization of advertising strategies using terms such as geotargeting, demographic targeting, personalized advertising, and A/B testing [8].

Goal: To enhance the efficiency of online advertising by optimizing advertising campaigns.

Tasks:

1. Comparison of existing methods for optimizing online advertising:
   - Analysis of modern approaches to optimizing online advertising campaigns.
   - Evaluation of their effectiveness and applicability.
2. Development of a probabilistic model for predicting advertising conversion using recurrent neural networks:
   - Creation of a mathematical model based on probabilistic methods for predicting customer conversions from advertising.
   - Validation of the model on real data.
3. Analysis and improvement of advertising budget management strategies.
   - Use of web analytics data to optimize the distribution of advertising budgets across different channels and ads.
   - Development of recommendations for improving advertising strategies.

2 Comparison of Existing Methods for Optimizing Online Advertising

Analysis of modern approaches: In the modern world, online advertising is a complex and dynamic sphere where effective strategies are constantly evolving. Several key methods have been identified through research into various approaches. Machine learning, including neural networks and supervised learning algorithms, takes a central position. Significant attention is also given to data analytics methods, including the use of probabilistic models. Figure 1 illustrates the predominant development of online advertising compared to offline, as well as the increase in revenue [1].

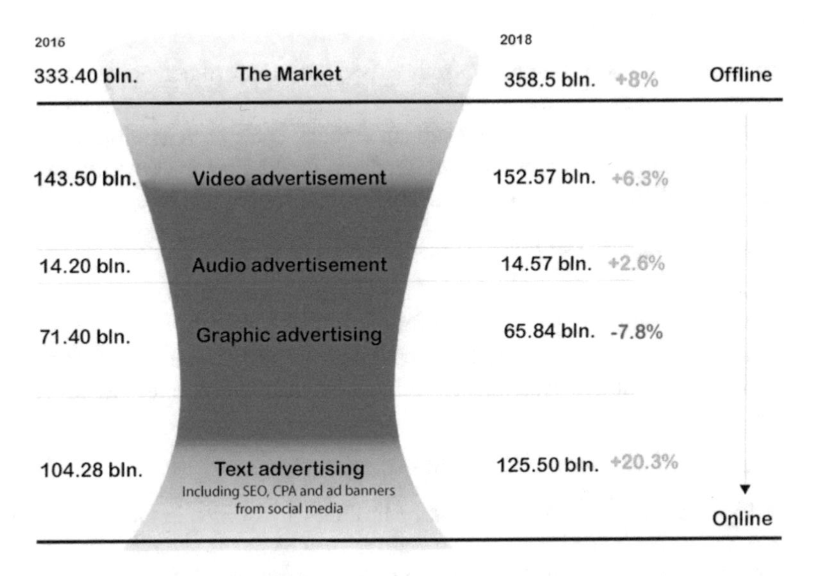

Fig. 1. Effectiveness of online advertisement.

Evaluation of Effectiveness: Machine learning methods such as recurrent neural networks (RNN) and deep learning algorithms demonstrate impressive efficiency in predicting conversion [6]. Nevertheless, they can demand a significant amount of data and resources. Although budget optimization methods and bid management algorithms offer stability in results, their accuracy may be lower.

Applicability to Different Scenarios: Machine learning methods are successfully applied in various fields but are particularly effective in campaigns with high dynamics and large datasets. On the other hand, traditional methods perform well in stable conditions and with limited budgets.

Resilience to Changes: Machine learning methods may exhibit resilience to changes, providing adaptive models. However, Traditional methods, on the other hand, tend to be more resistant to changes, whereas effectiveness of machine learning advertisement may diminish in the face of radical shifts in the market environment.

Examples of Successful Applications: Machine learning algorithms, such as deep learning models, have been successfully used by companies targeting the youth segment. Meanwhile, budget optimization methods prove effective in segments with high competition and limited resources [4].

Issues and Solutions for Existing Online Advertising Optimization Methods:

1. Ineffective Targeting:
   - Problem: Traditional methods may struggle with limited accuracy in identifying the target audience.
   - Solution: Models like Attention-based RNN can analyze multiple factors, allowing for more precise identification and prediction of user interests and preferences.
2. Lack of Personalization:
   - Problem: Advertising campaigns may experience a lack of personalized content or tailored messages, diminishing their effectiveness and hindering user engagement.
   - Solution: The attention mechanism in models, such as Attention-based RNN, enables highlighting individual preferences, facilitating the creation of personalized advertising messages.
3. Suboptimal Budget Allocation:
   - Problem: Traditional methods may face difficulties in optimizing the distribution of the advertising budget among different channels and ads.
   - Solution: Attention-based RNN models can optimize budget distribution to maximize return on investment (ROI) by dynamically adapting to changing conditions.

7. Limited Behavioral Analysis Capabilities:
   - Problem: Existing methods may have limited capabilities in analyzing complex patterns in user behavior.
   - Solution: Models using RNN and attention mechanisms can process sequential data and highlight key aspects of behavior, enhancing their analytical capabilities.

Conclusions from the Analysis: The analysis of existing methods highlights that the ideal advertising optimization strategy depends on the specific conditions and goals of the campaign. Machine learning methods offer greater flexibility but require attentive management. Target accuracy may be lacking in traditional methods, despite their stability. This conclusion serves as the foundation for selecting optimal strategies in the further development of the model.

3 Development of a Probabilistic Model Based on Web Analytics Using Attention-Based RNN and Model Evaluation

To address the task of predicting conversion in online advertising, we propose an effective solution based on Attention-based RNN (see Fig. 2).

This model has a unique ability of focusing on key aspects of data, making it an ideal tool for analyzing user behavior and optimizing advertising campaigns (see Fig. 3).

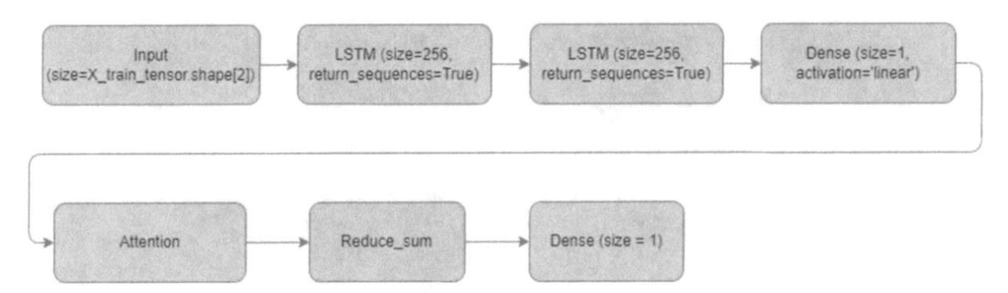

Fig. 2. Architecture of neural network.

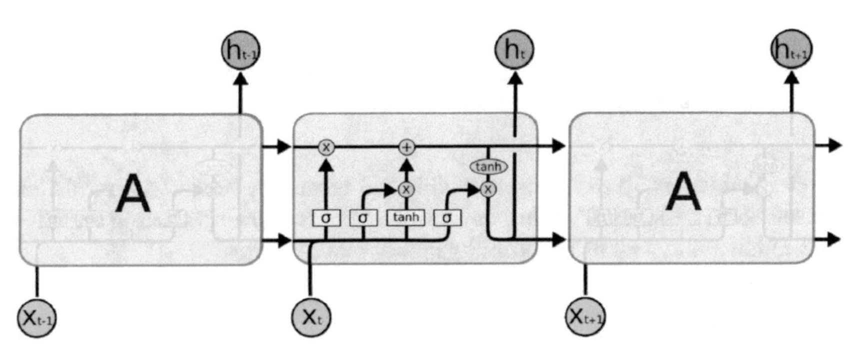

Fig. 3. Architecture of LMST.

Input layers: LSTM (Long Short-Term Memory).

LSTM (Long Short-Term Memory) is a type of recurrent neural network specifically designed to address the issue of gradient vanishing, which often occurs in simple RNNs. Instead of using a single activation function as in RNNs, LSTM employs special memory blocks, including input, output, and forget gates [12].

The use of LSTM in this model allows for considering long-term dependencies in website traffic data, which is a crucial factor for accurate prediction.

Attention Mechanism: The attention layer in our model plays a crucial role. Instead of uniformly averaging all time steps, as traditional RNNs do, attention enables the model to dynamically focus on specific moments in the data. In the context of a website, this can be particularly useful when visitor activity may vary during different periods (see Fig. 4).

Output Layer: Fully connected layer to generate the final output.

Model Capabilities:

1. Conversion Prediction: The model can predict the probability of conversion for each advertising campaign, considering various factors such as target audience, temporal parameters, and attention mechanism, enabling the creation of personalized advertising messages.
2. Budget Optimization: The model optimizes the distribution of the advertising budget across different channels and ads, maximizing Return on Investment (ROI).
3. A/B Testing: Capability to conduct A/B testing on different variations of advertising campaigns to identify the most successful approaches.

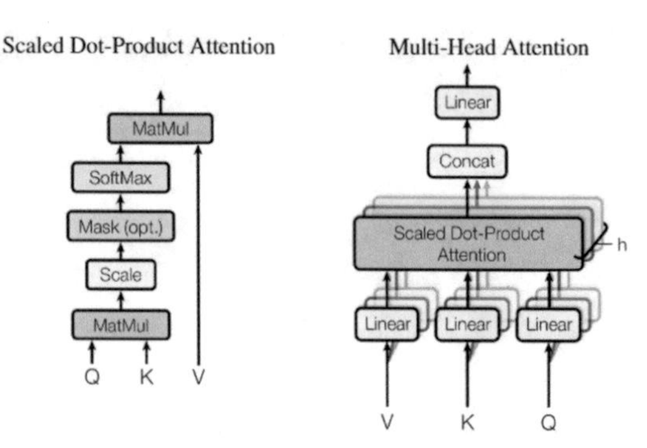

Fig. 4. On the left: attention based on scaled dot-product. On the right: "multi-headed" attention consists of multiple attention layers operating in parallel [13].

Model Validation: To ensure the reliability and accuracy of the model, validation on real-world data is essential. Testing on various datasets allows evaluating the model's generalization ability and its applicability in different scenarios.

3.1 Documentation

1. Initialization. Model parameters (input_size, hidden_size, output_size) need to be configured based on data characteristics.
2. Loss Function and Optimizer. Selected loss function - BCELoss (Binary Cross Entropy) for binary classification tasks. The optimizer is Adam with a learning rate of 0.001.
3. Training. The training process consists of iterating through a loop, during which the model processes input data, compares predictions to actual values, adjusts weights, and repeats these steps for multiple epochs.
4. Hyperparameter Tuning. Hyperparameter tuning was conducted to achieve optimal results. The following hyperparameters proved most effective:

   - hidden_size = 256: Size of the LSTM hidden layer, determining the number of neurons in each layer.
   - initial_learning_rate = 0.01: Initial learning rate, setting the weight update step at the beginning of training.
   - decay_steps = 3000: Number of steps after which the learning rate decreases.
   - decay_rate = 0.9: Coefficient for reducing the learning rate.
   - epochs = 10: Number of training epochs, determining how many times the model processes the entire training data.
   - batch_size = 32: Data batch size used for each weight update.

These hyperparameters were tuned with a balance between performance and training resources, allowing the model to effectively capture temporal dependencies in the data and achieve good results in prediction of website traffic.

5. Model Evaluation. After completing the model training, an evaluation is conducted on test data. The evaluation results include the Mean Absolute Error (MAE) and Mean

Squared Error (MSE). In this case, the MAE is 3.3, and the MSE is 13.4, indicating good model accuracy.

6. Mean Absolute Error (MAE). MAE represents the average absolute difference between actual and predicted values. Its choice is motivated by the following pros and cons:
   - Pros of MAE:
   - Interpretability
   - Robustness to outliers
   - Cons of MAE:
   - No consideration of weights
7. Mean Squared Error (MSE). MSE measures the average squared difference between actual and predicted values. The choice of MSE is also justified by a set of pros and cons:
   - Pros of MSE:
   - Sensitivity to large errors
   - Mathematical stability
   - Cons of MSE:
   - Sensitivity to outliers

The choice between MAE and MSE depends on the context of task and characteristics of data. MAE allows for simple interpretation and robustness to outliers, while MSE is more sensitive to large errors and mathematically more stable. Using both metrics in this work allows for a comprehensive understanding of the model's accuracy.

8. Conclusions. The Attention-based RNN model provides a powerful tool for analyzing and optimizing advertising campaigns by highlighting key aspects of data and predicting conversion. Model evaluation through testing, performance metrics, and application scenarios confirms its technical success and justifies business decisions in the online advertising domain.

4 Business Plan Development (Mathematical Model) - Project Development Strategy

Definition of Goals: The main goal of the business plan is to maximize the efficiency of online advertising using a probabilistic model based on web analytics and Attention-based RNN [7]. To achieve this goal, the following stages of project development need to be outlined.

1. Implementation Stage:
   - Goal. Launch the first version of the model and test it on a limited volume of data.
   - Steps.
   - Data Preparation. Integration of the model with data on advertising campaigns and users.
   - Testing and Debugging. Launching the model on a limited number of campaigns to identify potential issues and optimize.

- Feedback Collection. Obtaining feedback from users and advertisers for model adjustments and improvements.

2. Scaling Stage:

   - Goal. Expand the use of the model to a larger volume of data and advertising campaigns.
   - Steps.
   - Performance Optimization. Working on improving the performance and scalability of the model.
   - Data Expansion. Integration with a larger volume of diverse data for more accurate predictions.
   - Advertising Campaign. Launching a promotional campaign for the model to attract new clients.

3. Optimization Stage:

   - Goal. Improve accuracy and adapt the model to specific client needs.
   - Steps.
   - Data Analysis. In-depth analysis of data and identification of new factors to enhance the model.
   - Architecture Modification. Making changes to the model architecture based on the analysis results.
   - User Experience. Optimizing user experience and interaction with the platform.

4. Automation Stage:

   - Goal. Implement automation systems to simplify the use of the model and manage advertising campaigns.
   - Steps.
   - API Integration. Developing an API for convenient interaction with the model.
   - Management Systems. Creating a control panel for monitoring and managing advertising campaigns.
   - Client Training. Training clients on using the system and interpreting results.
   - Task for the Next Semester.
   - Product Effectiveness Verification/Evaluation.
   - Metric Definition. Refining key effectiveness metrics in the context of clients' business objectives.
   - Real Campaign Testing. Verifying the effectiveness of the model by implementing it into current advertising campaigns.

5. Conclusions. Each stage of model development will be accompanied by result analysis and adjustments to the strategy. A comprehensive conclusion will be drawn after completing the product effectiveness verification stage.

Figure 5 illustrates the effectiveness of increasing profit using online advertising. It can be concluded that with the use of the right algorithm, online advertising can be a key tool for generating profit [2].

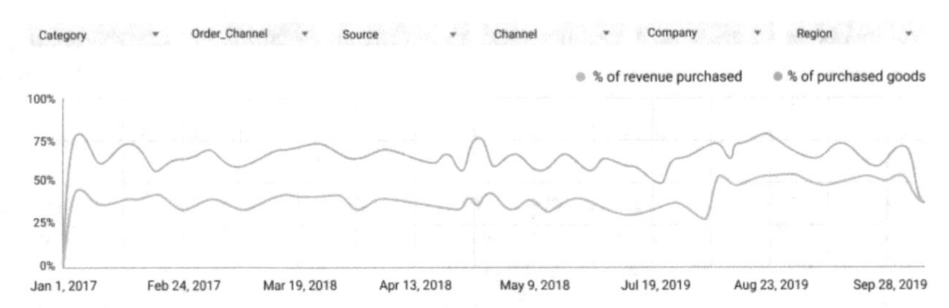

Fig. 5. Percentage breakdown of profit from online advertising.

5 Conclusion

The following tasks were accomplished during the project:

- An analysis and comparison of modern approaches to optimizing online advertising campaigns were conducted.
- A probabilistic model for predicting advertising conversion using recurrent neural networks was developed.
- Stages of project development were identified.

Enhancing the targeting of advertising campaigns, maximizing profits, and increasing conversions are the primary objectives of this project, which centers on optimizing online advertising through probabilistic models and web analytics.

References

1. Britvina, V.V., Zueva, A.S., Gavrilyuk, A.V., Petryaev, P.R., Shumakov, D.O.: Statistical and technical analysis of the development prospects of encrypted unregulated digital assets. In: Proceedings of SPIE - The International Society for Optical Engineering (2023)
2. Pyataeva, O., Britvina, V., Anisimov, A., Gavrilyuk, A., Nurgazina, G.: Analysis of statistical parameters of development and introduction of innovative technologies in branches of the Russian economy. In: E3S Web of Conferences (2023)
3. Pavlyuk, I.D., Britvina, V.V., Gavrilyuk, A.V., Niyazbekova, S.U., Nurgazina, G.E.: Using mathematical statistics to optimize the process of crossovers using data center infrastructure management. In: Proceedings of SPIE - The International Society for Optical Engineering (2023)
4. Samoylova, A.S., et al: The use of information technology and mathematical modeling in the development of modes of aluminum alloy. In: CEUR Workshop Proceedings (2021)
5. Samoylova, A., et al: Automation of calculation of the workload norms for engineers in the machine-building industry. In: Proceedings of SPIE - The International Society for Optical Engineering (2022)
6. Artificial Intelligence in IT: How Artificial Intelligence Will Transform the IT Industry. https://softengi.com/blog/ai-in-it-how-artificial-intelligence-will-transform-the-it-industry. Accessed 28 Oct 2023
7. How Artificial Intelligence, Machine Learning, and Deep Learning Work. https://trends.rbc.ru/trends/industry/5e845cec9a794747bf03e2c9. Accessed 28 Oct 2023
8. OWASP AI Security and Privacy Guide. https://owasp.org/www-project-ai-security-and-privacy-guide/. Accessed 28 Oct 2023

9. Features of using artificial neural networks in information security. https://cyberleninka.ru/article/n/osobennosti-ispolzovaniya-iskusstvennyh-neyronnyh-setey-v-sfere-informatsionnoy-bezopasnosti. Accessed 28 Oct 2023
10. Neural Networks in Cybersecurity. https://habr.com/ru/articles/587694/. Accessed 28 Oct 2023
11. Bogodukhova, E.S., Bobrova, E.O., et al.: Directions for the development of renewable energy sources in Russia using information technologies during the formation of the climate crisis. IOP Conf. Ser. Earth Environ. Sci. **723**(5) (2021)
12. Goodfellow, I., Bengio, Y., Courville, A.: Deep Learning. MIT Press, Massachusetts (2016)
13. Neural networks, graphs and emergence. https://habr.com/ru/articles/751340/. Accessed 28 Oct 2023

Development of a System for Modeling the Design and Optimization of the Operation of a Small Hydroelectric Power Station

G. N. Uzakov[1(✉)], Z. E. Kuziev[2], A. B. Safarov[1,2], and R. A. Mamedov[2]

[1] Karshi Engineering-Economics Institute, Karshi, Uzbekistan
uzoqov66@mail.ru

[2] Bukhara Engineering Technological Institute, Bukhara, Uzbekistan

Abstract. The work substantiates that small hydropower is actively developing in the world, as a result of which it seems relevant to develop a system for modeling the design and optimization of the operation of a small hydropower plant with the possibility of using various input parameters. When designing and modeling a hydropower plant, SolidWorks, ANSYS CFD, and MATLAB application packages were used. A mathematical model of the relationship between pressure and water flow, as well as the design parameters of a two-rotor hydroelectric power station, has been developed. By using the proposed modeling system, it is possible to design the construction of power plants and configure the optimal operating mode of hydroelectric power plants in any country in the world.

Keywords: Modelling · Vertical Axis Hydropower Plant · Ansys CFD · Polynomial Constants · Optimal Parameters

1 Introduction

In the world, one of the leading trends in energy is the use of renewable energy sources, especially at pumping stations, to save electricity and increase the reliability of power supply. "The World Energy Strategy plans to spend US$2.56 billion on small hydropower installations between 2021 and 2030 and increase utilization rates by 2.8%". In this regard, the development and implementation of microhydroelectric power plants with a vertical axis, adapted to low-pressure watercourses, is relevant [1].

Scientific research is being conducted around the world aimed at developing micro-hydroelectric power plants with a vertical axis of rotation, adapted to variable flow rates and pressures of water flow of pumping units, justifying mechanical, energy, hydrological and design parameters and increasing their efficiency. In this regard, the development of micro-hydroelectric power plants with a vertical axis of rotation, adapted to low-pressure watercourses, modeling of operating modes and design parameters, improving the method for increasing the efficiency of multi-pole magnetoelectric generators for hydropower plants, based on an algorithm, determining the optimal angle of inclination of hydraulic turbine blades, special attention is paid to the development of a mathematical

A. Gibadullin (Ed.): DITEM 2023, LNNS 942, pp. 243–252, 2024.
https://doi.org/10.1007/978-3-031-55349-3_20

model determining the optimal rotation speed and maximum useful efficiency of the impeller, substantiating their main energy parameters, as well as their implementation in practice [2, 3].

According to the International Energy Agency (IEA) and the International Renewable Energy Agency (IRENA), hydropower accounts for the largest share of renewable energy sources. At the end of 2022, the volume of electricity generated by hydroelectric power plants around the world amounted to more than 4.408 TWh (3.7% increase compared to 2021). Thus, hydropower accounts for 15% of global electricity production. In addition, hydropower has been increasingly used in recent years as one of the oldest and most reliable sources of renewable energy. More than 34 GW of new hydropower capacity was added globally in 2022, with total capacity increasing to 1.397 GW, up 2.7% from 2021. For the first time since 2018, more than 30 GW of new capacity was commissioned in 2022. Figure 1 shows the dynamics of hydropower growth in the world in 2018–2022. [4, 5].

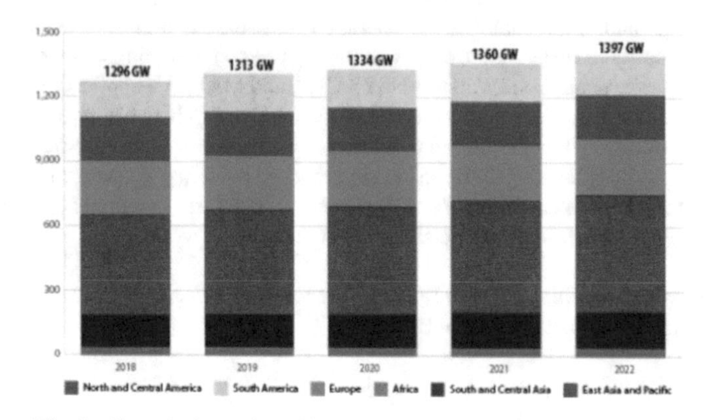

Fig. 1. Growth dynamics of hydropower in the world in 2018–2022.

Research is underway on the development and implementation of low-power installations based on renewable energy sources, adapted to the climatic conditions of Uzbekistan [6–8]. Let's analyze scientific research carried out in the world and in Uzbekistan on the development of hydropower plants with a vertical axis of rotation to increase their efficiency in low-pressure watercourses.

Polish scientist D. Borkowski, in his research work, developed a model of a microhydropower installation operating with variable water flow. The study used a propeller-type hydraulic turbine with a guide vane, and a magnetoelectric synchronous generator was used as an electric generator. In the research work, the nominal efficiency coefficient was determined to be $\eta = 0.75$, while the angle of entry of the water flow into the impeller with a radius of 0.33 m is 148°, the nominal water flow is 3.5 m^3/s, the nominal water pressure is 3 m [9].

Russian scientists V.M. Ivanov in their research, evaluated the possibilities of using guiding apparatus and hydroturbines in micro hydropower devices that can be used in horizontal and vertical water flows. Expressions for determining the structural dimensions of 5–10 kW hydropower plants when the pressure of the water flow is 5–10 m and the water consumption is 75–85 l/s are presented and analyzed [10].

Lakhdar Belhadji et al., in a research work, designed a micro-hydroelectric power plant operating on variable water flows. This installation consists of a device that directs the flow of water, a propeller-type (semi-Kaplan) hydraulic turbine and an electric generator. A graph is presented of the dependence of the efficiency on the angular velocity at various values of water flow in the micro-hydroelectric power station being developed. Based on this graph, it was established that the turbine efficiency is in the range of 65–85% with a water flow rate of 0.11–0.18 m^3/s and a change in angular velocity in the range of 0–2700 rpm [11].

A.B. Mamajanov in his research, developed a gravity-vortex microhydroelectric power station. At this station, due to inverted conical blades with a water flow pressure inside a cylindrical pool of 1.7 m and a water flow rate of 1.466 m^3/s, an increase in the installation power to 17.6 kW and an efficiency of 85% was achieved [12].

O.O. Bazarov in his dissertation, created a microhydroelectric power station with a reactive hydraulic unit to supply electricity to agriculture. Technical parameters of this device: water pressure 2 m, water flow 200 l/s and power 2.35 kW. This hydroelectric power station uses an asynchronous generator [13].

D.B. Kodirov, in his doctoral dissertation, developed a system for the combined use of solar and water energy for dispersed rural and water energy consumers. His work proposed a solar-hydroelectric power plant taking into account the volume and speed of water flow, the power and time of solar radiation. This 4.5 kW installation is designed to generate an average of 0.84–1.39 kWh of electricity using solar and water energy, taking into account various weather conditions [14].

Despite the success achieved, for water flows from pumping units with variable flow rate and pressure of water flow with a water flow direction device, a multi-pole advanced magnetoelectric generator is placed in a special basin where the water flow rotates for uniform distribution. Transfer of water flows from the pipeline of a pumping unit to the blades of the device, and research on the development and justification of the parameters of a micro-hydroelectric power station with a vertical axis installed at an optimal angle in order to reduce hydraulic resistance forces has not been sufficiently studied.

Purpose of the study: justification of parameters by modeling and optimization of a microhydroelectric power station with a vertical axis, adapted to the water flows of pumping units and operating effectively in low-pressure watercourses.

2 Methods and Results

Hydropower plant 1 operates as follows (Fig. 2). A metal frame 3 is installed above the water flow of the irrigation channel 2, on which a stationary vertical shaft 5 with a hydraulic turbine is mounted, driven into rotation by the flow of water coming out of the pipe 4 of the pumping unit. When the water flow interacts with the pool 7 in which the guide vane 8 is installed, the water flow receives rotational motion. As a result of the passage of water flow through the guide vane 8, a uniform rotation of the upper impeller 10 begins clockwise. The flow of water flowing from the blades of the upper impeller

10 falls on the guide vanes 17 mounted on a fixed shaft 5, which is designed to direct the flow of water to the lower impeller 13 to rotate it in a counterclockwise direction. More precisely, the impeller, through a belt drive 28, rotates a metal disk 26, which is connected to the armature of the electric generator using bushings 27. As a result, the armature 22 of the generator 24 receives a clockwise rotating motion. The lower impeller, through a belt drive 29, rotates the flange joint connected to the inductors 19,20 of the electric generator by means of bolts. As a result, the generator inductor consisting of permanent magnets 21 receives a counterclockwise rotating motion. The voltage generated on the three-phase winding 23 of the armature 22 of the generator is transmitted through the cable to the brushes 30 with slip rings 31. As a result of the opposite rotation of the armature 22 and the inductors 19,20 of the axial magnetoelectric generator 24 in opposite directions, high electromagnetic power is achieved [15].

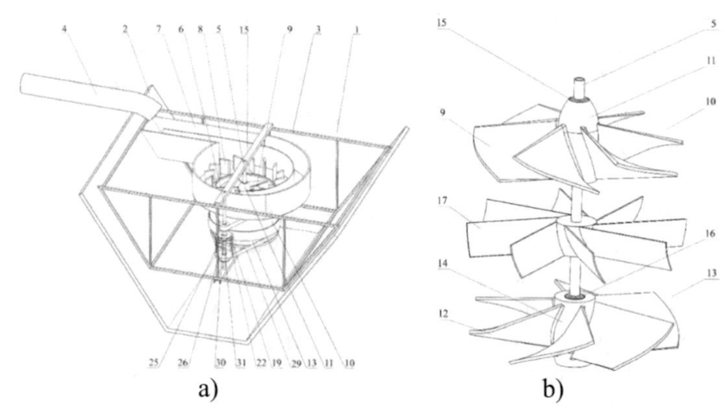

Fig. 2. a) appearance of the developed hydropower installation; b) installation blades and guide surface.

Figure 3 shows a model of hydraulic turbine blades developed in the SolidWorks program. The velocity triangle was used to determine the optimal blade angle.

In Fig. 4 shows the results of the model built in Ansys CFD and Solidworks application package to determine the mechanical parameters of the developed vertical axis hydraulic turbine. At the same time, when assessing the rotation speed of the hydraulic turbine blades, the change in pressure, speed and flow rate of the water flow was taken into account.

A mathematical model of the dependence of the energy indicators of a hydropower plant with a vertical axis on changes in pressure and flow rate of water flow is as follows [16–18].

Equation for the dependence of water pressure on turbine rotation speed and water flow:

$$H(Q_i, n) = A_1 \cdot Q^m + B_1 \cdot Q \cdot n + C_1 \cdot n^2 \tag{1}$$

where: A_1, B_2 and C_2 are polynomial constants depending on pressure, water flow and turbine rotation speed; Q_i - water flow rate, m^3/s; n – rotation speed of the hydraulic turbine, rpm.

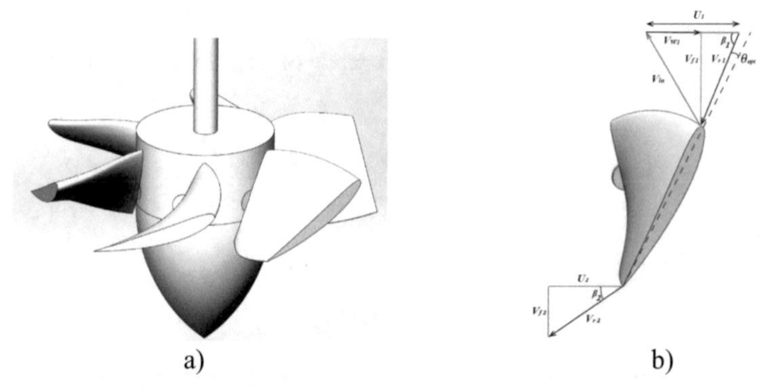

a) b)

Fig. 3. Shows a model of hydraulic turbine blades developed in the SolidWorks program. The velocity triangle was used to determine the optimal blade angle.

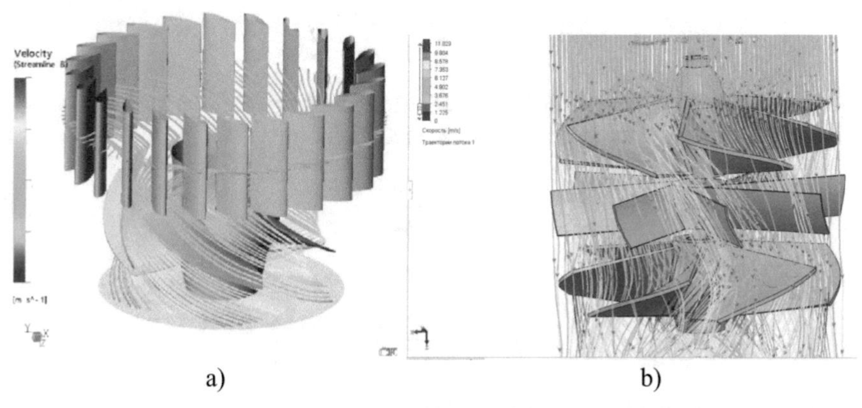

a) b)

Fig. 4. Dynamics of changes in water flow in an Ansys CFD hydraulic turbine (a), dynamics of changes in Solidworks water flow rate.

Mathematical expression for the mechanical power of a hydraulic turbine:

$$P_m(Q_i, n) = A_3 \cdot Q \cdot n^2 + B_3 \cdot n \cdot Q^2 \tag{2}$$

where: A_3 and B_3 are polynomial constants depending on water flow, mechanical power and turbine rotation speed.

Mechanical torque of the hydraulic turbine:

$$M_m(Q_i, n) = A_2 \cdot Q^2 + B_2 \cdot Q \cdot n \tag{3}$$

where: A_2 and B_2 are polynomial constants of mechanical torque depending on water flow and turbine rotation speed.

The dependence of the efficiency factor on water flow and rotation speed is found from the ratio of the mechanical power of the hydraulic turbine and the power of water flow [19].

$$\eta(Q_i, n) = \frac{P_m(Q_i, n)}{\rho \cdot g \cdot H(Q_i, n) \cdot Q_i} \tag{4}$$

where: ρ - water density, kg/m^3; g- free fall acceleration, m/s^2

Expression of the dependence of the efficiency of a hydraulic turbine on water flow and rotation speed:

$$\eta(Q_i, n) = \frac{A_3 \cdot Q_i \cdot n^2 + B_3 \cdot n \cdot Q_i^2}{\rho \cdot g \cdot \left(A_1 \cdot Q_i^3 + B_1 \cdot Q_i^2 \cdot n + C_1 \cdot Q_i \cdot n^2\right)} \tag{5}$$

Hub diameter:

$$d_{hub} = m \cdot d_{run} \tag{6}$$

where: d_{run} – turbine diameter, m; m – represents the relationship between diameters and is determined depending on the pressure of the water flow.

Water flow rate [20, 21]:

$$V_{f1} = \frac{Q_i}{S} = \frac{Q_i}{\frac{\pi}{4} \cdot \left(d_{run}^2 - d_{hub}^2\right)} \tag{7}$$

where: S – area of the blades affected by the water flow, m^2.

Tangential speed of the hydraulic turbine [22]:

$$U_1 = \frac{\pi \cdot d_{run} \cdot n}{60} \tag{8}$$

Linear speed of hydraulic turbine [24]:

$$V_{wi} = \frac{P_m \cdot 1000}{\rho \cdot Q_i \cdot U_1} \tag{9}$$

Angle of water supply to the blades [23]:

$$\beta_1 = \text{arc tg}\frac{V_{f1}}{U_1 - V_{w1}} \tag{10}$$

Angle of water flow from the blades:

$$\beta_2 = \text{arc tg}\frac{V_{f1}}{U_1} \tag{11}$$

Optimal blade angle [24, 25]:

$$\theta = 180^{\circ} - \beta_1 + \alpha \tag{12}$$

A, B and C – constants depend on the rotation speed and power of the turbine. Therefore, the characteristics of the dependence are presented by rotation speed and water flow (Table 1). Ansys CFD application package was used to determine polynomial constants.

Figure 5 shows the dependences of the water flow pressure (a), mechanical torque (b), mechanical power (c) and efficiency (d) of a hydraulic turbine at a water flow rate of 6–12 m^3/s and a hydraulic turbine rotation speed of 0 ... 150 rpm In Fig. 5 (d), from the

Table 1. Polynomial constants.

A1	B1	C1	A2	B2	m	A3	B3
0.022	0.0054	0.0006	0.5023	0.0589	1.763	0.0062	0.0526

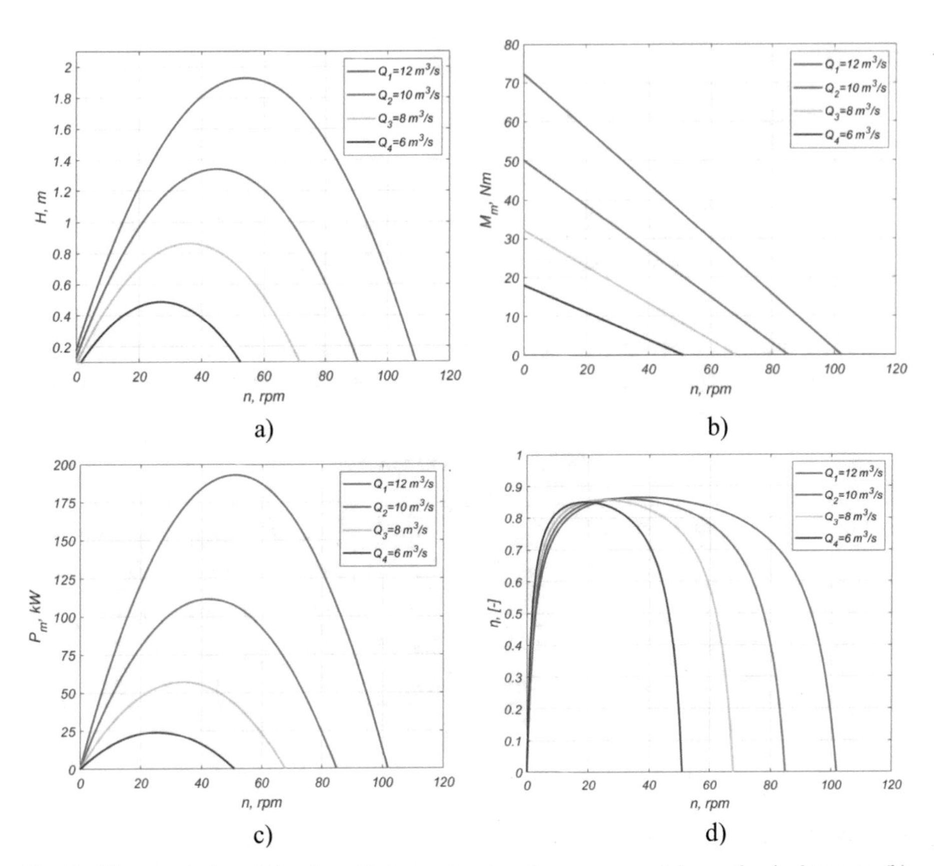

Fig. 5. Characteristics of the dependences of water flow pressure (a), mechanical torque (b), mechanical power (c) and efficiency factor (d) at various water flow rates on the rotation speed of the hydraulic turbine.

graphs of the dependence of the efficiency on the rotation speed at various water flow rates, it is established that the value of the efficiency at the optimal rotation speed of the hydraulic turbine is 0.88.

In Fig. 6 shows the dependences of water flow speed, consumption and pressure changes on the optimal angle of inclination of the hydraulic turbine, rotation speed and outer diameter of the hydraulic turbine modeled in the Matlab system. In this case, the input power parameters are 25–190 kW, water pressure 0.4–2 m, water flow 6–12 m^3/s, maximum efficiency 0.88, number of blades 6 and tilt angle water flow interacting with the blades of the hydraulic turbine 50.

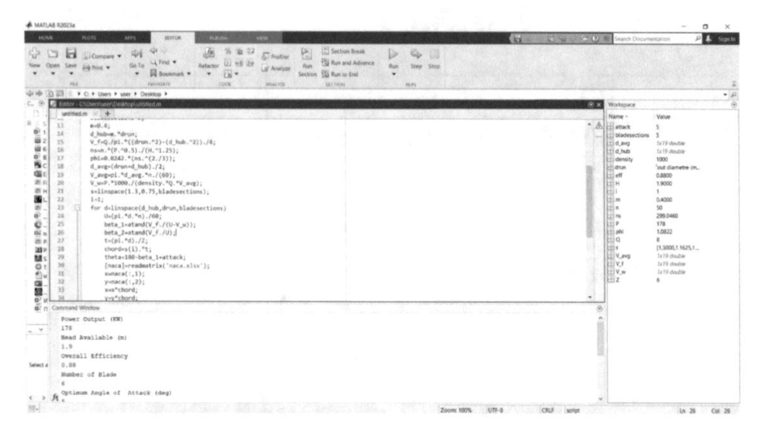

Fig. 6. Modelling of hydraulic turbines in Matlab.

Table 2 presents the results of determining the optimal parameters of the proposed hydropower plant depending on the variable pressure and flow rate of the water flow.

Table 2. Optimal parameters.

No	Q, m^3/s	n_{opt}, rpm	P_{opt}, kW	V_{f1}, m/s	U_1, m/s	V_{w1}, m/s	β1, grad	θopt. Grad
1	6	25	25	9.1	1.3	24.2	−22	207
2	8	35	60	12.1	1.8	7.9	−64	248
3	10	45	115	15.2	2.3	2.6	−89	274
4	12	55	190	18.1	2.8	0.7	83	102

3 Conclusion

Analysis of studies on the use of water energy in the world and in Uzbekistan showed that this is an important strategic direction in reducing the negative consequences of the energy crisis, efficient use of energy resources, stabilizing environmental problems, developing the social sphere, economic sectors and solving their problems. A mathematical model has been developed for the dependence of the pressure and flow rate of water flow, as well as the design parameters of a two-rotor hydropower plant, adapted to the water flow of pumping units. In this case, the design dimensions of the installation are, outer diameter 1 m, inner diameter 0.4 m, number of blades 6, number of blades directing the water flow 16, diameter of the water pool 1.5 m, diameter of the pool 1.5 m, water flow rate 6–12 m^3/s, the optimal speed of the hydraulic turbine at a head of 0.5–2 m is 25–55 rpm, mechanical torque10–60 Nm, mechanical power 25–190 kW, optimal blade installation angles are 102°–207° degrees and maximum efficiency values are 0.85–0.88. This hydropower plant is capable of producing 1.66 GWh of electricity per year when operating at full capacity. Through the use of the proposed hydropower plants, it is possible to achieve increased energy efficiency of farms and pumping stations located far from the centralized energy supply.

References

1. Fernandez, L.: Global small hydropower market value forecast 2021–2030. https://www.statista.com/statistics/790736/global-small-hydropower-market-size/. Accessed 11 Oct 2023
2. Chamil, A.: Modelling and optimisation of a Kaplan turbine - A comprehensive theoretical and CFD study. Clean. Energy Syst. **3**, 100017 (2022). https://doi.org/10.1016/j.cles.2022.100017
3. Ardizzon, G., Cavazzini, G., Pavesi, G.: A new generation of small hydro and pumped hydro power plants: advances and future challenges. Renew. Sustain. Energy Rev. 746–761 (2014). https://doi.org/10.1016/j.rser.2013.12.043
4. 2023 World Hydropower Outlook. https://indd.adobe.com/view/92d02b04-975f-4556-9cfe-ce90cd2cb0dc. Accessed 11 Oct 2023
5. IRENA (2023), World Energy Transitions Outlook 2023: 1.5°C Pathway, Volume 2, International Renewable Energy Agency, Abu Dhabi. https://www.irena.org/Publications/2023/Jun/World-Energy-Transitions-Outlook-2023. Accessed 11 Oct 2023
6. Sadullaev, N.N., Safarov, A.B., Mamedov, R.A., Kodirov, D.: Assessment of wind and hydropower potential of Bukhara region. IOP Conf. Ser. Earth Environ. Sci. **614**(1), 012036 (2020)
7. Safarov, A., Davlonov, H., Mamedov, R., Chariyeva, M., Kodirov, D.: Design and modeling of dynamic modes of low speed electric generators for electric power generation from renewable energy sources In: AIP Conference Proceedings, vol. 2686, p. 020013 (2022)
8. Safarov, A.B., Mamedov, R.A.: Study of effective Omni-directional vertical axis wind turbine for low speed regions IIUM. Eng. J. **22**(2), 149–160 (2022)
9. Dariusz, B.: Analytical model of small hydropower plant working at variable speed. IEEE Trans. Energy Convers. **10**, 1109 (2018). https://doi.org/10.1109/tec.2018.2849573
10. Ivanov, V.M., Ivanova, T.Y., Jdanov, Y.P., Kleyn, G.O., Yurenkov, V.N.: Methodology for calculating the flow path of an axial hydraulic turbine of a new original design. Polzunovsky Bulletin **4**, 253–258 (2009)
11. Lakhdar, B., Seddik, B., Iulian, M., Axel, R., Daniel, R.: Adaptive MPPT applied to variable-speed microhydropower plant. IEEE Trans. Energy Convers. **22** (2012). https://doi.org/10.1109/tec.2012.2220776
12. Mamadzhanov, A.B.: Improving the energy efficiency of the gravity vortex micro hydroelectric power station. PhD thesis, Tashkent (2023)
13. Bazarov, O.O.: Creation of a micro-hydroelectric unit with reactive hydro aggregate for agricultural consumers. PhD thesis, Tashkent (2020)
14. Kadirov, D.B.: A systematic approach to the use of renewable energy in rural and water supply. DSc thesis, Tashkent (2022)
15. Ministry of Justice of the Republic of Uzbekistan "Vertical axis hydropower device, IAP 07462, 17.08.2023" patent
16. Samora, I., Hasmatuchi, V., Münch-Alligne, C., et al.: Experimental characterization of a five blade tubular propeller turbine for pipe inline installation. Renew. Energy **95**, 356–366 (2016)
17. Jawahar, C.P., Michael, P.A.: A review on turbines for micro hydro power plant. Renew. Sustain. Energy Rev. **72**, 882–887 (2017)
18. Singh, P., Nestmann, F.: Experimental optimization of a free vortex propeller runner for micro hydro application. Exp. Therm. Fluid Sci. **33**, 991–1002 (2009)

19. Dariusz, B., Marek, M.: Small hydropower plants with variable speed operation-an optimal operation curve determination. Energies **13**, 6230 (2020). https://doi.org/10.3390/en13236230
20. Valavi, M., Nysveen, A.: Variable-speed operation of hydropower plants: past, present, and future. In: 22nd International Conference on Electrical Machines, Lausanne, pp. 640–646 (2016)
21. Pospehov, G.B., Savón, Y., Delgado, R., Castellanos, E.A., Peña, A.: Inventory of landslides triggered by hurricane Matthews in Guantánamo, Cuba. Geogr. Environ. Sustain. **16**(1), 55–63 (2023). https://doi.org/10.24057/2071-9388-2022-133
22. Yang, W., Yang, J.: Advantage of variable-speed pumped storage plants for mitigating wind power variations: Integrated modelling and performance assessment. Appl. Energy **237**, 720–732 (2019). https://doi.org/10.1016/j.apenergy.2018.12.090
23. Raguzin, I.I., Bykowa, E.N., Lepikhina, O.U.: Polygonal metric grid method for estimating the cadastral value of land plots. Lomonosov Geogr. J. **78**(3), 92–103 (2023). https://doi.org/10.55959/MSU0579-9414.5.78.3.8
24. Fraile-Ardanuy, J., Wilhelmi, J.R., Fraile-Mora, J.J., Pérez, J.I.: Variable-speed hydro generation: operational aspects and control. IEEE Trans. Energy Convers. **21**, 569–574 (2006)
25. Bykowa, E., Skachkova, M., Raguzin, I., Dyachkova, I., Boltov, M.: Automation of negative infrastructural externalities assessment methods to determine the cost of land resources based on the development of a thin client model. Sustainability **14**, 9383 (2022). https://doi.org/10.3390/su14159383
26. Valavi, M., Nysveen, A.: Variable-speed operation of hydropower plants: a look at the past, present and future. IEEE Ind. Appl. Mag. **24**, 18–27 (2018). https://doi.org/10.1109/mias2017.2740467
27. Pospehov, G.B., Savon-Vaciano, Y., Hernandez-Columbie, T.: Landslide processes as a natural disturbance in ecosystems in the Alejandro de Humboldt National Park, Cuba. IOP Conf. Ser. Earth Environ. Sci. **1212**, 012029 (2023). https://doi.org/10.1088/1755-1315/1212/1/012029
28. Uzakov, G.N., Kuziev, Z.E.: Study of constructive dimensions of Kaplan hydro turbine in variable water flows. Spectr. J. Innov. Reforms Dev. **20**, 36–42 (2023)
29. Dong, Z., et al.: Developing of quaternary pumped storage hydropower for dynamic studies. IEEE Trans. Sustain. Energy **11**, 2870–2878 (2020). https://doi.org/10.1109/tste.2020.2980585

Author Index

A. Gibadullin (Ed.): DITEM 2023, LNNS 942, pp. 253–254, 2024.
https://doi.org/10.1007/978-3-031-55349-3

Printed in the United States
by Baker & Taylor Publisher Services